On Maps from Loop Suspensions to Loop Spaces and the Shuffle Relations on the Cohen Groups

MEMOIRS
of the
American Mathematical Society

Number 851

On Maps from Loop Suspensions
to Loop Spaces and the Shuffle
Relations on the Cohen Groups

Jie Wu

March 2006 • Volume 180 • Number 851 (end of volume) • ISSN 0065-9266

American Mathematical Society
Providence, Rhode Island

2000 *Mathematics Subject Classification.*
Primary 55Pxx; Secondary 18G20, 20C05, 55Qxx, 55Uxx.

Library of Congress Cataloging-in-Publication Data

Wu, Jie, 1964–
   On maps from loop suspensions to loop spaces and the shuffle relations on the Cohen groups / Jie Wu.
       p. cm. — (Memoirs of the American Mathematical Society, ISSN 0065-9266 ; no. 851)
   "Volume 180, number 851 (end of volume)."
   Includes bibliographical references.
   ISBN 0-8218-3875-X (alk. paper)
     1. Homotopy theory. 2. Algebra, Homological. 3. Representations of groups. 4. Algebraic topology. I. Title. II. Series.
QA3.A57   no. 851
[QA612.7]
510 s—dc22
[514′.24]   2005058190

Memoirs of the American Mathematical Society

This journal is devoted entirely to research in pure and applied mathematics.

Subscription information. The 2006 subscription begins with volume 179 and consists of six mailings, each containing one or more numbers. Subscription prices for 2006 are US$624 list, US$499 institutional member. A late charge of 10% of the subscription price will be imposed on orders received from nonmembers after January 1 of the subscription year. Subscribers outside the United States and India must pay a postage surcharge of US$31; subscribers in India must pay a postage surcharge of US$43. Expedited delivery to destinations in North America US$35; elsewhere US$130. Each number may be ordered separately; *please specify number* when ordering an individual number. For prices and titles of recently released numbers, see the New Publications sections of the *Notices of the American Mathematical Society*.

Back number information. For back issues see the *AMS Catalog of Publications*.

Subscriptions and orders should be addressed to the American Mathematical Society, P. O. Box 845904, Boston, MA 02284-5904, USA. *All orders must be accompanied by payment.* Other correspondence should be addressed to 201 Charles Street, Providence, RI 02904-2294, USA.

Copying and reprinting. Individual readers of this publication, and nonprofit libraries acting for them, are permitted to make fair use of the material, such as to copy a chapter for use in teaching or research. Permission is granted to quote brief passages from this publication in reviews, provided the customary acknowledgment of the source is given.

Republication, systematic copying, or multiple reproduction of any material in this publication is permitted only under license from the American Mathematical Society. Requests for such permission should be addressed to the Acquisitions Department, American Mathematical Society, 201 Charles Street, Providence, Rhode Island 02904-2294, USA. Requests can also be made by e-mail to `reprint-permission@ams.org`.

Memoirs of the American Mathematical Society is published bimonthly (each volume consisting usually of more than one number) by the American Mathematical Society at 201 Charles Street, Providence, RI 02904-2294, USA. Periodicals postage paid at Providence, RI. Postmaster: Send address changes to Memoirs, American Mathematical Society, 201 Charles Street, Providence, RI 02904-2294, USA.

© 2006 by the American Mathematical Society. All rights reserved.
Copyright of this publication reverts to the public domain 28 years
after publication. Contact the AMS for copyright status.
This publication is indexed in *Science Citation Index*®, *SciSearch*®, *Research Alert*®,
CompuMath Citation Index®, *Current Contents*®/*Physical, Chemical & Earth Sciences*.
Printed in the United States of America.

∞ The paper used in this book is acid-free and falls within the guidelines
established to ensure permanence and durability.
Visit the AMS home page at `http://www.ams.org/`

10 9 8 7 6 5 4 3 2 1   11 10 09 08 07 06

Contents

Introduction	1
Chapter 1. Maps from Loop Suspensions to Loop Spaces	9
1.1. The James Construction	9
1.2. Bi-Δ-groups	16
1.3. Skeletons of Bi-Δ-groups	26
1.4. The Cohen Construction	31
Chapter 2. Shuffle Relations	39
2.1. Functors to Coalgebras	39
2.2. Geometric Realizations and the Shuffle Relations	53
2.3. Proof of Theorem 3	56
2.4. Proof of Theorem 4	58
Bibliography	63

Abstract

The maps from loop suspensions to loop spaces are investigated using group representations in this article. The shuffle relations on the Cohen groups are given. By using these relations, a universal ring for functorial self maps of double loop spaces of double suspensions is given. Moreover the obstructions to the classical exponent problem in homotopy theory are displayed in the extension groups of the dual of the important symmetric group modules $\mathrm{Lie}(n)$, as well as in the top cohomology of the Artin braid groups with coefficients in the top homology of the Artin pure braid groups.

Received by the editor January 25, 2005.

2000 *Mathematics Subject Classification*. Primary 55P; Secondary 18G20, 20C05,55Q, 55U.

Key words and phrases. loop spaces, James construction, Hopf invariants, bi-Δ-groups, Cohen groups, functors to coalgebras, shuffle relations.

Research is supported in part by the Academic Research Fund of the National University of Singapore.

Introduction

[1]In [**9, 10, 11**], Cohen introduced a combinatorial group $\mathfrak{H}$ with a representation to the group of the homotopy classes of functorial self maps of loop suspensions. This group is important for studying the classical exponent problem in homotopy theory, for instance the classical results in [**2**] can be obtained by some simple combinatorial computations in the group $\mathfrak{H}$. In [**39, 40, 41, 45, 48**], the group $\mathfrak{H}$ has been successfully applied to solve some problems in the classical homotopy theory including the Cohen conjecture. As a combinatorial group, $\mathfrak{H}$ has connections with homotopy string links studied by Milnor and Habegger-Lin in low dimensional topology [**27, 28, 19, 23**], as well as braid groups and simplicial groups [**3, 13, 46**].

The purpose of this article is to study the maps from loop suspensions to loop spaces using group representations. In particular, we obtain a generalization of the Cohen group $\mathfrak{H}$. Our generalization provides a way to construct various Cohen type groups related to maps from loop suspensions to loop spaces. By investigating the reduced diagonals, we obtain the shuffle relations on the Cohen groups. The quotient of the Cohen group $\mathfrak{H}$ by the shuffle relations gives a universal ring $\mathfrak{R}$ for functorial self maps of double loop spaces of double suspensions. The ring $\mathfrak{R}$ is related to the extension groups of the important symmetric group modules Lie(n) by investigating functors to coalgebras. Moreover the obstructions to the exponent problem in homotopy theory are displayed in these extension groups. In addition to homotopy theory, the representation theory of the ring $\mathfrak{R}$ is related to the functorial version of the Poincaré-Birkhoff-Witt Theorem, with connections to the modular representation theory of the symmetric groups. The remainder of this introduction describes our results in more details.

Let X be a pointed space. Recall that the *James construction* $J(X)$ is the free monoid generated by points in X modulo the single relation that the basepoint $* = 1$, with the weak topology. The classical James theorem [**21**] states that $J(X)$ is (weak) homotopy equivalent to $\Omega\Sigma X$ if X is path-connected. The *James filtration* $\{J_n(X)\}$ is the word filtration, namely $J_n(X)$ is the quotient space of X^n by the equivalence relation generated by

$$(x_1, \ldots, x_{i-1}, *, x_i, \ldots, x_{n-1}) \sim (x_1, \ldots, x_{j-1}, *, x_j, \ldots, x_{n-1})$$

for any $2 \leq i, j \leq n$. Let $X^{(n)}$ denote the n-fold self smash product of X. Then $J_n(X)/J_{n-1}(X)$ is homeomorphic to $X^{(n)}$. Let $q_n \colon X^n \to J_n(X)$ be the quotient map. By the suspension splitting theorem [**21**], the inclusions $J_{n-1}(X) \subseteq J_n(X)$ induce a tower of group epimorphisms

$$[J(X), \Omega Y] \twoheadrightarrow \cdots \twoheadrightarrow [J_n(X), \Omega Y] \twoheadrightarrow \cdots \twoheadrightarrow [X, \Omega Y]$$

---

[1]Received by the editor January 25, 2005.

Research is supported in part by the Academic Research Fund of the National University of Singapore.

with the inverse limit $[J(X), \Omega Y] = \lim_n [J_n(X), \Omega Y]$ and group monomorphism

$$q_n^* \colon [J_n(X), \Omega Y] \rightarrowtail [X^n, \Omega Y]$$

for each $n \geq 1$. Our study on the group of homotopy classes $[J(X), \Omega Y]$ is given by introducing operations on the sequence of groups $\{[X^{n+1}, \Omega Y]\}_{n \geq 0}$ described as follows. Note that the group $[J(X), \Omega Y] \cong [\Omega \Sigma X, \Omega Y]$ if X is a path-connected CW-complex.

The coordinate inclusions and projections

$$d_i \colon X^{n+1} \to X^n \quad (x_0, x_1, \ldots, x_n) \mapsto (x_0, \ldots, x_{i-1}, x_{i+1}, \ldots, x_n)$$

$$d^i \colon X^n \to X^{n+1} \quad (x_0, x_1, \ldots, x_{n-1}) \mapsto (x_0, \ldots, x_{i-1}, *, x_i, \ldots, x_{n-1})$$

induce functions (group homomorphisms if Z is a homotopy associative H-space with inverse)

$$d_i = d^{i*} \colon [X^{n+1}, Z] \to [X^n, Z] \quad \text{and} \quad d^i = d_i^* \colon [X^n, Z] \to [X^{n+1}, Z]$$

for any pointed space Z with the following identities:

$$d_j d_i = d_i d_{j+1} \quad \text{for} \quad j \geq i,$$

$$d^j d^i = d^{i+1} d^j \quad \text{for} \quad j \leq i,$$

$$d_j d^i = \begin{cases} d^{i-1} d_j & \text{for} \quad j < i, \\ \text{id} & \text{for} \quad j = i, \\ d^i d_{j-1} & \text{for} \quad j > i, \end{cases}$$

where $d_0 x = *$ for $x \in [X, Z]$. The sequence of sets $\mathfrak{K}(X, Z) = \{[X^{n+1}, Z]\}_{n \geq 0}$ with the operations d_i and d^i has the property that $\mathfrak{K}(X, Z)$ is a Δ-set under the faces given by d_i and a co-Δ-set under the cofaces given by d^i with the third identity mixing faces and cofaces together. The third identity is different from simplicial identities. Motivated from the above three identities, we introduce the concept of *bi-Δ-set* (*bi-Δ-group*), namely, a sequence of sets (groups) with faces (face homomorphisms) and cofaces (coface homomorphisms) satisfying the above identities. The face operations are important for understanding $[J(X), Z]$ by the following lemma. On the other hand, coface operations are used for constructing Hopf invariants.

Let $\mathcal{S} = \{S_n\}_{n \geq 0}$ be a Δ-set, the *Cohen set* (*Cohen group* if $\mathcal{S}$ is a Δ-group) $\mathfrak{H}_n \mathcal{S}$ is defined to be the equalizer of the faces, namely

$$\mathfrak{H}_n \mathcal{S} = \{x \in S_n \mid d_0 x = d_1 x = \cdots = d_n x\}.$$

The *total Cohen set* (*total Cohen group* if $\mathcal{S}$ is a Δ-group) is defined to be the inverse limit

$$\mathfrak{H} \mathcal{S} = \lim_{p_n} \mathfrak{H}_n \mathcal{S},$$

where $p_n = d_0|_{\mathfrak{H}_n \mathcal{S}} \colon \mathfrak{H}_n \mathcal{S} \to \mathfrak{H}_{n-1} \mathcal{S}$. (**Note.** By the identity on faces,

$$d_i|_{\mathfrak{H}_n \mathcal{S}} = d_0|_{\mathfrak{H}_n \mathcal{S}}$$

for each $0 \leq i \leq n$.)

LEMMA 1 (Theorem 1.1.5). *Let X and Z be path-connected spaces. Suppose that X is a co-H-space or Z is an H-space. Then $[J_{n+1}(X), Z] \cong \mathfrak{H}_n \mathfrak{K}(X, Z)$ and $[J(X), Z] \cong \mathfrak{H} \mathfrak{K}(X, Z)$.*

In other words, under the above assumptions, the set of homotopy classes $[J(X), Z]$ can be determined by the sets $[X^{n+1}, Z]$ together with face operations and so simplicial techniques can be applied for studying $[J(X), Z]$. For instance, the Moore-Postnikov system (Theorem 1.2.7) gives a bi-Δ-group resolution of $\mathfrak{K}(X, \Omega Y)$. In the universal cases, the Moore-Postnikov system coincides with the descending central series by Proposition 1.4.6.

By applying the $\mathfrak{H}$-construction to simplicial groups, we have the following surprising results:

THEOREM 2 (Theorem 1.2.2). *Let* $\mathcal{G} = \{G_n\}_{n \geq 0}$ *be a reduced simplicial group, that is,* $G_0 = \{1\}$. *Then*
1) $p_{2k+1} \colon \mathfrak{H}_{2k+1}\mathcal{G} \to \mathfrak{H}_{2k}\mathcal{G}$ *is an epimorphism for each* $k \geq 0$.
2) *The image of* $p_{2k} \colon \mathfrak{H}_{2k}\mathcal{G} \to \mathfrak{H}_{2k-1}\mathcal{G}$ *is a normal subgroup of* $\mathfrak{H}_{2k-1}\mathcal{G}$ *with cokernel isomorphic to* $\pi_{2k-1}(\mathcal{G})$ *for each* $k \geq 1$.

It was conjectured by Fred Cohen that $\mathfrak{H}\mathcal{G}$ is a progroup for any simplicial group G, that is $\mathfrak{H}_n\mathcal{G} \to \mathfrak{H}_{n-1}\mathcal{G}$ is always onto. Our answer is that $\mathfrak{H}\mathcal{G}$ is a progroup if and only if $\mathcal{G}$ has trivial odd dimensional homotopy groups. Moreover the cokernels of $\mathfrak{H}_n\mathcal{G} \to \mathfrak{H}_{n-1}\mathcal{G}$ are given by the odd dimensional homotopy groups of $\mathcal{G}$. This gives systematic method for obtaining all odd dimensional homotopy groups and killing off all even dimensional homotopy groups.

When a Δ-group $\mathcal{G}$ admits a bi-Δ-structure, $\mathfrak{H}\mathcal{G}$ is always a progroup according to Proposition 1.2.1. Moreover Theorem 1.2.4 gives the Taylor series for decomposing elements in the Cohen group $\mathfrak{H}\mathcal{G}$ of a bi-Δ-group $\mathcal{G}$ in terms of other elements. This result is a reformulation of the classical distributivity law [**1, 5, 8**] in the case of $\mathfrak{K}(X, \Omega Y)$ and has applications to the exponent problem. As an example, Example 1.2.8 gives the bi-Δ-group structure on pure braids and so the Taylor series gives certain canonical decomposition of certain type of braids.

For generalizing the results in [**9, 10, 11**], we investigate the skeleton filtration of bi-Δ-groups that can be regarded as the categorical interpretation of the skeleton filtration of CW-complexes or the dual version of the Moore-Postnikov systems. Roughly speaking a skeleton of a bi-Δ group $\mathcal{G}$ is obtained by taking the lower dimensional groups from $\mathcal{G}$ (including faces and cofaces) and "freeing up" higher dimensional groups to form a bi-Δ-group. It admits the universal property described as in Theorem 1.4.10. This gives another bi-Δ-group resolution of $\mathfrak{K}(X, \Omega Y)$. In particular, the 0-skeleton gives the universal group for $[\Omega\Sigma X, \Omega\Sigma X]$ for X running over all path-connected spaces. According to the remarks to Corollary 1.3.9, this is the progroup built up by the Brunnian braids, where a braid is called *Brunnian* if it becomes trivial by removing any one of its strings. When X runs over all path-connected p-local spaces, the p-local version of the universal group is canonically obtained. Moreover various local type of the universal group can be obtained by allowing X to run over certain class of spaces such as localization with respect to a homology theory or a class of maps. According to Proposition 1.3.7, the 0-skeleton is obtained by self free products of a given group with certain canonical faces and cofaces.

In addition to the 0-skeleton, it should be pointed out that the higher skeleton of $\mathfrak{K}(X, \Omega Y)$ is also important for individual spaces X and Y. For instance, the divisibility of certain elements occurs in the 1-skeleton of $\mathfrak{K}(S^n, \Omega Y)$ if the Whitehead square ω_{n+1} is divisible by 2. It is a long-standing open problem whether the Whitehead square ω_{2^k-1} is divisible by 2. Another example is the bi-Δ-group $\mathfrak{K}(S^1, \Omega Y)$.

As a sequence of groups, this is the bi-Δ-group obtained from $[S^1 \times \cdots \times S^1, \Omega Y]$ with its Cohen group given by $[\Omega S^2, \Omega Y]$ that is isomorphic to $\prod_{n=1}^{\infty} \pi_n(\Omega Y)$ as a set. The skeleton filtration on $\mathfrak{K}(S^1, \Omega Y)$ is obtained by taking the lower homotopy groups of ΩY and so it is related to the Postnikov system of ΩY.

The commutator relations in [**9, 10, 11**] can be generalized by considering group homomorphisms $\phi\colon G \to [X, \Omega Y]$. A group homomorphism $\phi\colon G \to [X, \Omega Y]$ is called to have *weak LS-category* less than k if the composite

$$X \xrightarrow{\bar{\Delta}_k} X^{(k)} \xrightarrow{[[\phi(g_1),\phi(g_2)],\ldots,\phi(g_k)]} \Omega Y$$

is null homotopic for any $g_1, g_2, \ldots, g_k \in G$, where $X^{(k)}$ is the k-fold self-smash product of X, $\bar{\Delta}_k$ is the reduced diagonal and $[[\phi(g_1), \phi(g_2)], \ldots, \phi(g_k)]$ is the iterated Samelson product of the maps $\phi(g_1), \ldots, \phi(g_k)$. For instance, any group homomorphism $\phi\colon G \to [X, \Omega Y]$ has weak LS-category less than k if X has weak LS-category less than k (that is, the reduced diagonal $\bar{\Delta}_k\colon X \to X^{(k)}$ is null homotopic). By using weak LS-category filtration, the universal group admits a resolution by adding certain systematic type of commutator relations. According to Theorem 1.4.10, as the special case that $k = 1$, the resulting quotient group extends the main results in [**9, 10, 11**]. Roughly speaking, assuming that X is a path-connected co-H-space with a given subgroup G in $[X, \Omega Y]$, then there is a progroup with a group homomorphism to $[\Omega \Sigma X, \Omega Y]$ that is built up by $\mathrm{Lie}(n)$ with coefficients in self tensor products of G, where $\mathrm{Lie}(n)$ will be described below. (See Section 1.4 for details.)

The group $\mathrm{Lie}(n)$ is constructed as follows. Let $\bar{V}$ be a free $\mathbb{Z}$-module of rank n with a basis $\{x_1, x_2, \ldots, x_n\}$. The module $\mathrm{Lie}(n)$ is defined to be the submodule of $\bar{V}^{\otimes n}$ spanned by the iterated commutators

$$[[x_{\sigma(1)}, x_{\sigma(2)}], \ldots, x_{\sigma(n)}]$$

for $\sigma \in S_n$, where $[x, y] = xy - yx$. (**Note.** For $n = 1$, $\mathrm{Lie}(1)$ is defined to $\mathbb{Z}$.) The symmetric group S_n acts on $\mathrm{Lie}(n)$ by permuting the letters $\{x_1, \ldots, x_n\}$. It is well known that $\mathrm{Lie}(n)$ is a free $\mathbb{Z}$-module of rank $(n-1)!$, see for instance [**7, 35**]. For any abelian R, write $\mathrm{Lie}^R(n)$ for $\mathrm{Lie}(n) \otimes_{\mathbb{Z}} R$. Note that $\mathrm{Lie}^R(n)$ is a module over the symmetric group algebra $R(S_n)$ if R is a ring. The applications of the symmetric group module $\mathrm{Lie}^R(n)$ can be found in [**39, 40**]. Roughly speaking the problem on functorial decompositions of the $\Omega\Sigma X$ is equivalent to the problem for determining the maximal projective submodule of $\mathrm{Lie}^R(n)$ over the symmetric group. Our results described below give further connections between the homological properties of $\mathrm{Lie}^R(n)$ over the symmetric group and the possible obstructions to the classical exponent problem.

In [**9**], Cohen constructed a progroup $\mathfrak{H}$ built by $\mathrm{Lie}(n)$, that is, there is a tower of group epimorphisms

$$\cdots \twoheadrightarrow \mathfrak{H}_n \twoheadrightarrow \mathfrak{H}_{n-1} \twoheadrightarrow \cdots \twoheadrightarrow \mathfrak{H}_1 = \mathbb{Z}$$

such that $\mathrm{Ker}(\mathfrak{H}_n \twoheadrightarrow \mathfrak{H}_{n-1}) = \mathrm{Lie}(n)$ and the group $\mathfrak{H}$ is given by the inverse limit $\mathfrak{H} = \lim_n \mathfrak{H}_n$. Similarly there is a progroup $\mathfrak{H}^R$ built up by $\mathrm{Lie}(n)^R$ using our terminology of bi-Δ-groups. It was then proved in [**9**] that the group $\mathfrak{H}$ is a subgroup of (homotopy) self natural transformations of the functor $\Omega\Sigma^2$, that is, $\mathfrak{H}$ admits a functorially faithful representation $e_X\colon \mathfrak{H} \to [\Omega\Sigma^2 X, \Omega\Sigma^2 X]$. Moreover

there is a commutative diagram

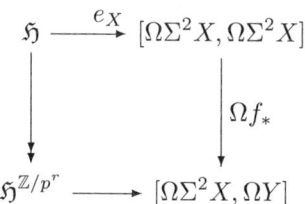

for any map $f\colon \Sigma^2 X \to Y$ of order p^r in $[\Sigma^2 X, Y]$. Following the lines in [9] the progroup $\mathfrak{H}^{\mathbb{Z}_{(p)}}$ ($\mathfrak{H}^{\mathbb{Z}_p}$), built up by $\mathrm{Lie}(n)$ over p-local integers $\mathbb{Z}_{(p)}$ (p-complete integers $\mathbb{Z}_p$), is a subgroup of (homotopy) self natural transformations of $\Omega\Sigma^2$ on p-local (p-complete) spaces.

It was discovered in [39, 40] that the progroup $\mathfrak{H}^R$ is isomorphic to the group of functorial self coalgebra maps of tensor algebras. More precisely, let R be a given ground ring and let $T(V)$ be the tensor algebra generated by a projective module V with the canonical Hopf algebra structure by saying V primitive. Consider $T\colon V \mapsto T(V)$ as a functor from projective R-modules to pointed cocommutative coalgebras. Then the set of natural (coalgebra) self transformations of T forms a group under the the convolution product. Let $\mathrm{coalg}^{R,\mathrm{graded}}(T,T)$ and $\mathrm{coalg}^R(T,T)$ denote the groups of functorial self coalgebra maps of tensor algebras over graded modules and ungraded modules, respectively. Write $[\Omega\Sigma^2, \Omega\Sigma^2]$ for the group of homotopy natural transformations of the functor $\Omega\Sigma^2$. (**Note.** Following Quillen's ideas, for avoiding set theoretical troubles, $\Omega\Sigma^2$ is regarded as a functor from CW-complexes with cells indexed by a fixed (large) infinite set to pointed spaces.) Let $\mathfrak{H}^R$ be the progroup constructed in [9, 10, 11] that is built up by $\mathrm{Lie}(n)$ with coefficients in R. By using Geometric Realization Theorem [39, Theorem 1.3], the composite

$$\mathfrak{H}^R \xrightarrow{e} [\Omega\Sigma^2, \Omega\Sigma^2] \xrightarrow{H_*} \mathrm{coalg}^{R,\mathrm{graded}}(T,T) \to \mathrm{coalg}^R(T,T) \cong \mathfrak{H}^R$$

is an isomorphism of groups for $R = \mathbb{Z}, \mathbb{Z}_{(p)}, \mathbb{Z}_p$, where H_* is the homology functor with coefficients in R. As a consequence, it follows that $\mathfrak{H}^R$ occurs as a splitting subgroup of $[\Omega\Sigma^2, \Omega\Sigma^2]$.

Let $T(V) \wedge T(V)$ be the quotient module of $T(V) \otimes T(V)$ by $T_i(V) \otimes_R R \oplus R \otimes T_i(V)$ for $i > 0$. Let $\bar{\psi}$ be the reduced comultiplication given by the composite $\bar{\psi}\colon T(V) \xrightarrow{\psi} T(V) \otimes T(V) \twoheadrightarrow T(V) \wedge T(V)$. Then $\bar{\psi}$ is a functorial coalgebra map and so it induces a group homomorphism

$$\bar{\psi}^*\colon \mathrm{coalg}^R(T \wedge T, T) \longrightarrow \mathrm{coalg}^R(T,T) \cong \mathfrak{H}^R$$

with cokernel denoted by $\mathfrak{R}^R$. Note that the explicit formula for the $\bar{\psi}$ is given by the shuffles. The group $\mathfrak{R}^R$ can be regarded as the quotient of $\mathfrak{H}^R$ by certain shuffle type relations.

Our next result gives a relation between $\mathfrak{R}^R$ and the group homomorphism

$$\Omega\colon [\Omega\Sigma^2, \Omega\Sigma^2] \to [\Omega^2\Sigma^2, \Omega^2\Sigma^2] \quad [f] \mapsto [\Omega f].$$

Let $\mathrm{Lie}^R(n)^*$ denote the dual module of $\mathrm{Lie}^R(n)$.

THEOREM 3. *Let R be a commutative ring with identity. Then there is a quotient group $\mathfrak{R}^R$ of $\mathfrak{H}^R = \mathrm{coalg}^R(T,T)$ with the following properties.*

1) $\mathfrak{R}^R$ is an abelian group. Moreover $\mathfrak{R}^R$ is a ring with the multiplication induced by the composition operation on $\mathrm{coalg}^R(T,T)$.
2) For any ring homomorphism $\phi\colon R\to R'$, there is an induced ring homomorphism
$$\mathfrak{R}(\phi)\colon \mathfrak{R}^R \to \mathfrak{R}^{R'}$$
for changing the group rings, that is, $\mathfrak{R}$ defines a functor from commutative rings with identity to rings with identity.
3) There is a morphism of rings
$$\theta\colon \mathfrak{R}^R \longrightarrow \prod_{n=1}^\infty \mathrm{Hom}_{R(S_n)}(\mathrm{Lie}^R(n),\mathrm{Lie}^R(n)).$$
If R is a field of characteristic 0, then θ is an isomorphism.
4) There is a tower of epimorphisms of rings
$$\cdots \twoheadrightarrow \mathfrak{R}_n^R \twoheadrightarrow \mathfrak{R}_{n-1}^R \twoheadrightarrow \cdots \twoheadrightarrow \mathfrak{R}_1^R = R$$
such that $\mathfrak{R}^R = \lim_n \mathfrak{R}_n^R$.
5) Let $\mathfrak{I}_n^R$ denote the kernel of $\mathfrak{R}_n^R \to \mathfrak{R}_{n-1}^R$. Then there is an exact sequence
$$0 \longrightarrow \mathrm{Ext}_{R(S_n)}(\mathrm{Lie}^R(n)^*, \mathrm{Lie}^R(n)^*) \longrightarrow \mathfrak{I}_n^R \xrightarrow{\theta} \mathrm{Hom}_{R(S_n)}(\mathrm{Lie}^R(n),\mathrm{Lie}^R(n))$$
$$\longrightarrow H^0_{R(S_n)}(\mathrm{Lie}^R(n)^*;\mathrm{Lie}^R(n)^*) \longrightarrow 0$$
for each n.
6) If $R=\mathbb{Z}_{(p)}$ (or $\mathbb{Z}_p$), there is a commutative diagram of semirings

$$\begin{array}{ccccc}
\mathfrak{H}^R & \longrightarrow & \mathfrak{R}^R & \xrightarrow{\theta} & \prod_{n=1}^\infty \mathrm{Hom}_{R(S_n)}(\mathrm{Lie}^R(n),\mathrm{Lie}^R(n)) \\
\downarrow {\scriptstyle e} & & \downarrow {\scriptstyle e} & & \uparrow \\
[\Omega\Sigma^2,\Omega\Sigma^2] & \xrightarrow{\Omega} & [\Omega^2\Sigma^2,\Omega^2\Sigma^2] & \xrightarrow{\Omega^{k-2}} & [\Omega^k\Sigma^2,\Omega^k\Sigma^2]
\end{array}$$

for each $k\geq 2$, where $\Omega^k\Sigma^2$ is regarded as a functor from p-local (or p-complete) spaces to pointed spaces. Moreover if $f\colon \Sigma^2 X\to Y$ is of order p^r in $[\Sigma^2 X, Y]$, then there is a commutative diagram

$$\begin{array}{ccc}
\mathfrak{R}^R & \xrightarrow{e_X} & [\Omega^2\Sigma^2 X, \Omega^2\Sigma^2 X] \\
\downarrow & & \downarrow {\scriptstyle \Omega^2 f_*} \\
\mathfrak{R}^{\mathbb{Z}/p^r} & \longrightarrow & [\Omega^2\Sigma^2 X, \Omega^2 Y].
\end{array}$$

Remark. The map $\theta\colon \mathfrak{H}^R \to \prod_{n=1}^\infty \mathrm{Hom}_{R(S_n)}(\mathrm{Lie}(n),\mathrm{Lie}(n))$ is obtained by restricting coalgebra maps to primitives. If $R=\mathbb{Z}/p^r$, then, for each $t\geq r$, the convolution p^t-th power of tensor algebras (that corresponds to the p-th power map of the loop suspensions, also to $p^t\cdot \mathrm{id}$ in $\mathfrak{R}^R$) lies in $\mathrm{Ker}(\theta)$ that gives obstructions in the extension group of $\mathrm{Lie}(n)^*$ by part (5). So roughly speaking the (possible) obstructions to the powers of $\Omega^2 f$ are displayed in the extension groups

$\text{Ext}_{\mathbb{Z}/p^r(S_n)}(\text{Lie}^{\mathbb{Z}/p^r}(n)^*, \text{Lie}^{\mathbb{Z}/p^r}(n)^*)$. So far it is not clear whether the representation $e\colon \mathfrak{R}^R \to [\Omega^2\Sigma^2, \Omega^2\Sigma^2]$ is faithful or not for $R = \mathbb{Z}_{(p)}$ although by Part (3) of Theorem 3 this is faithful when $R = \mathbb{Q}$. By changing the ground ring $\mathbb{Z}_{(p)}$ to be $\mathbb{Q}$, $\mathfrak{R}^{\mathbb{Z}_{(p)}} \to \mathfrak{R}^{\mathbb{Q}}$ has the kernel that are detected by the extension groups of $\text{Lie}(n)^*$ and so the possible kernel of the representation e in $\mathfrak{R}_n^{\mathbb{Z}_{(p)}}$ level is a torsion group. This seems to suggest that there are certain geometric version of the Bocksteins.

This project is related to the classical exponent problem and so we give some historical remarks on the exponent problem. Let p be a prime integer and let G be an abelian group. Write $\text{Tor}_p(G)$ for the p-torsion component of G. The power p^r is called an *exponent* of G if $p^r \cdot \text{Tor}_p(G) = 0$. Let X be a space. The integer p^r is called an exponent of $\pi_*(X)$ if $p^r \cdot \text{Tor}_p(\pi_n(X)) = 0$ for all $n \geq 2$. If X is a simply connected CW complex of finite type, then each $\pi_n(X)$ has a bounded exponent which depends on n in general. For instance $\pi_*(S^2 \vee S^2)$ does not have a bounded exponent. On the other hand, one can check that the homotopy groups of the mapping space from a simply connected finite torsion space to a space has a bounded exponent.

It was first known by James [22] that $\pi_*(S^{2n+1})$ has an exponent bounded by 2^{2n}. The improvements given in [8, 38] state that the exponent of $\pi_*(S^{2n+1})$ is bounded by $2^{2n-[n/2]}$, where $[a]$ is the maximal integer $\leq a$. In the cases where $p > 2$, Toda [44] showed that $\pi_*(S^{2n+1})$ has an exponent bounded by p^{2n}. Selick [37] then showed that $\pi_*(S^3)$ has an exponent bounded by p for $p > 2$. Later Cohen-Moore-Neisendorfer [12] proved that $\pi_*(S^{2n+1})$ has an exponent bounded by p^n for $p > 2$. This is the best exponent because a theorem of Gray [18] gives that there are $\mathbb{Z}/p^n$-summands in $\pi_*(S^{2n+1})$ for $p > 2$.

The Barratt conjecture states that if $f\colon \Sigma^2 X \to Y$ is of order p^r in the group $[\Sigma^2 X, Y]$, then $p^{r+1} \cdot \text{Im}(f_*\colon \pi_*(\Sigma^2 X) \to \pi_*(Y)) = 0$; in particular, if the identity map $\text{id}_{\Sigma^2 X}$ is of order p^r in $[\Sigma^2 X, \Sigma^2 X]$, then $p^{r+1}\pi_*(\Sigma^2 X) = 0$. It was known by Neisendorfer [31] that the Barratt conjecture holds for (the identity map of) the Moore spaces $P^n(p^r)$ with $p > 2$, but it remains open in general.

Cohen proposed a strong form of the Barratt conjecture that if $f\colon \Sigma^2 X \to Y$ is of order p^r in the group $[\Sigma^2 X, Y]$, then $\Omega^2 f\colon \Omega^2\Sigma^2 X \to \Omega^2 Y$ has an order bounded by p^{r+1} in the group $[\Omega^2\Sigma^2 X, \Omega^2 Y]$. Note that if this statement is true, then the Barratt conjecture holds for f. Following Cohen's ideas there are essentially two steps for the study of this problem. Firstly study the group $[\Omega\Sigma^2 X, \Omega Y]$ for decomposing the powers of $[\Omega f]$ as a product of other type of maps, and secondly investigate the group homomorphism $\Omega\colon [\Omega\Sigma^2 X, \Omega Y] \to [\Omega^2\Sigma^2 X, \Omega^2 Y]$ for hoping to kill out the obstructions to the powers of $[\Omega f]$ after looping.

The first combinatorial study on the group $[\Omega\Sigma^2 X, \Omega Y]$ was given by Fred Cohen [9, 10, 11], where the group $\mathfrak{H}$ was introduced. Our results give more systematic investigation on the group $[\Omega\Sigma X, \Omega Y]$. Roughly speaking, we can start with an arbitrary subgroup of $[X, \Omega Y]$ for constructing a universal group for $[\Omega\Sigma X, \Omega Y]$, rather than starting with a cyclic subgroup generated by a homotopy class $[f] \in [X, \Omega Y]$. This generalization provides a possible way for collecting more information for studying $[\Omega\Sigma X, \Omega Y]$. Moreover our terminology of bi-Δ-groups provides a useful tool for connecting the universal group and the individual group $[\Omega\Sigma X, \Omega Y]$ by adding possible relations and possible new elements. The examples in [48] show that it is important to make connections between the universal group and the individual group $[\Omega\Sigma X, \Omega Y]$, where some special properties of the mod 2 Moore spaces are used.

For studying the homomorphism $\Omega\colon [\Omega\Sigma^2 X, \Omega Y] \to [\Omega^2\Sigma^2 X, \Omega^2 Y]$, Theorem 3 is essentially obtained by considering the reduced diagonal $\bar\Delta\colon \Omega\Sigma^2 X \to (\Omega\Sigma^2 X) \wedge (\Omega\Sigma^2 X)$ and using the well-known result that $\Omega\bar\Delta$ is null homotopic. The identification of $\mathfrak{H}$ with functorial coalgebra self maps of tensor algebras gives a useful information for computational purpose using shuffles. It seems that these are the first type of canonical relations by applying the cobar resolution to $\Omega\Sigma^2 X$.

The obstruction groups $\mathfrak{I}_n^R$ can be also described as the top cohomology of the Artin braid group B_n with coefficients in $\mathrm{Lie}(n)$ by the following theorem. For an $R(S_n)$-module M, write $M[-1]$ for the module M with signed S_n-action, that is, $M[-1] = M \otimes R[-1]$.

THEOREM 4. *Let $\mathfrak{I}_n^R$ be given in Theorem 3 and let B_n act on $\mathrm{Lie}^R(n)$ via the canonical quotient $B_n \to S_n$. Then there is an isomorphism*
$$\mathfrak{I}_n^R \cong H^{n-1}(B_n; \mathrm{Lie}^R(n)[-1]) \cong H^{n-1}(B_n; H_{n-1}(P_n; R))$$
where $P_n = \mathrm{Ker}(B_n \to S_n)$ is the Artin pure braid group with the canonical B_n-action on $H_*(P_n; R)$.

This theorem gives some connections with the complexity of algorithms studied by Smale [**42**] and other people [**15, 16, 17**].

The article is organized as follows. In Section 1.1, we go over the James construction, where Lemma 1 is Theorem 1.1.5. Bi-Δ-groups are introduced in Section 1.2, where some basic properties such as the Taylor series (Theorem 1.2.4) and the Postnikov system (Theorem 1.2.7) are given. Theorem 2 is Theorem 1.2.2. In Section 1.3, a categorical skeleton filtration of bi-Δ-groups is given. The generalized Cohen construction is given in Section 1.4. In Section 2.1, we investigate functors to coalgebras. The geometric realizations are discussed in Section 2.2 and the proof of Theorem 3 is given in Section 2.3. The proof of Theorem 4 is given in Section 2.4.

The author is indebted to Professors Jon Berrick, Fred Cohen, Bill Dwyer, John Harper and Paul Selick for fruitful conversations on this project. These conversations defined what was most important in this paper. The author would like to thank the referee for his/her important comments and helpful suggestions. The author also would like to thank many of our colleagues for their encouragement and suggestions on this project.

CHAPTER 1

Maps from Loop Suspensions to Loop Spaces

1.1. The James Construction

In this section, we go over the James construction. Some of the results in this section may be well-known.

Let X be a pointed space and let $J(X)$ be the James construction with the James filtration $\{J_n(X)\}$. For a subspace A of X, write $(X|_k A)^n$ for the subspace of X^n consisting of points $(x_1, x_2, \ldots, x_n) \in X^n$ with at least k coordinates lie in A. We simply write $(X|A)^n$ for $(X|_1 A)^n$. Note that $X^n / (X|*)^n = X^{(n)}$ and $(X|_{n-1}*)^n = \bigvee_{j=1}^n X$. Let $q_n \colon X^n \to J_n(X)$ be the canonical quotient map. Then there is a natural commutative diagram

(1.1.1)
$$\begin{array}{ccccccc} (X|_{n-1}*)^n = \bigvee_{j=1}^n X & \subseteq & \cdots & \subseteq & (X|_1 *)^n & \subseteq & X^n \\ \downarrow & & & & \downarrow & & \downarrow q_n \\ J_1(X) = X & \subseteq & \cdots & \subseteq & J_{n-1}(X) & \subseteq & J_n(X) \end{array}$$

for any (pointed) space X. In particular, there is a commutative diagram of cofibre sequences

$$\begin{array}{ccccc} (CX|X)^n & \hookrightarrow & (CX)^n & \twoheadrightarrow & (\Sigma X)^{(n)} \\ \downarrow \partial_n & & \downarrow & & \parallel \\ J_{n-1}(\Sigma X) & \hookrightarrow & J_n(\Sigma X) & \twoheadrightarrow & (\Sigma X)^{(n)}, \end{array}$$

where $CX = X \wedge [0,1]$ is the cone of X. The map

$$\partial_n \colon (CX|X)^n \to J_{n-1}(\Sigma X)$$

is called the *higher Whitehead product*, studied in [32, 33, 34]. Since $(CX)^n$ is contractible, we have the following:

PROPOSITION 1.1.1. *Let X be any pointed space. There is a (functorial) cofibre sequence*

$$(CX|X)^n \simeq \Sigma^{n-1} X^{(n)} \xrightarrow{\partial_n} J_{n-1}(\Sigma X) \longrightarrow J_n(\Sigma X).$$

Thus the cofibre sequence

$$J_{n-1}(\Sigma X) \hookrightarrow J_n(\Sigma X) \twoheadrightarrow (\Sigma X)^{(n)}$$

is principal. □

COROLLARY 1.1.2. *Let X be any pointed space. There is a (functorial) co-action of $(\Sigma X)^{(n)}$ on $J_n(\Sigma X)$, $\mu'\colon J_n(\Sigma X) \longrightarrow J_n(\Sigma X) \vee (\Sigma X)^{(n)}$, such that the diagram of cofibre sequences*

$$
\begin{array}{ccccc}
J_{n-1}(\Sigma X) & \longrightarrow & J_n(\Sigma X) & \longrightarrow & (\Sigma X)^{(n)} \\
\Big\| & & \Big\downarrow \mu' & & \Big\downarrow \\
J_{n-1}(\Sigma X) & \longrightarrow & J_n(\Sigma X) \vee (\Sigma X)^{(n)} & \longrightarrow & (\Sigma X)^{(n)} \vee (\Sigma X)^{(n)}
\end{array}
$$

commutes functorially. □

From the classical suspension splitting theorem [21], $q_n^*\colon [J_n(X), Y] \to [X^n, Y]$ is a monomorphism if Y is a loop space. This result can be generalized as follows:

PROPOSITION 1.1.3. *Let X and Y be pointed spaces. Then the kernel of the function*

$$q_n^*\colon [J_n(X), Y] \longrightarrow [X^n, Y]$$

is trivial, that is, if $f\colon J_n(X) \to Y$ such that $f \circ q_n$ is null homotopic, then so is f.

PROOF. The proof is given by induction on n. Clearly the statement holds for $n = 1$. Suppose that the statement holds for $n-1$ with $n \geq 2$. Let $f\colon J_n(X) \to Y$ be any map such that $f \circ q_n$ is null homotopic. From the commutative diagram

$$
\begin{array}{ccc}
X^{n-1} & \hookrightarrow & X^n \\
\Big\downarrow q_{n-1} & & \Big\downarrow q_n \\
J_{n-1}(X) & \hookrightarrow & J_n(X),
\end{array}
$$

the map f restricted to $J_{n-1}(X)$ is null homotopic and hence there is a map $g\colon X^{(n)} \to Y$ such that f is homotopic to the composite

$$J_n(X) \xrightarrow{p_n} X^{(n)} = J_n(X)/J_{n-1}(X) \xrightarrow{g} Y.$$

Consider the commutative diagram of cofibre sequence

$$
\begin{array}{ccccccc}
J_{n-1}(X) & \hookrightarrow & J_n(X) & \xrightarrow{p_n} & X^{(n)} & \longrightarrow & \Sigma J_{n-1}(X) \\
\Big\uparrow & & \Big\uparrow q_n & & \Big\| & & \Big\uparrow \\
(X|*)^n & \hookrightarrow & X^n & \xrightarrow{p_n \circ q_n} & X^{(n)} & \xrightarrow{\partial} & \Sigma(X|*)^n.
\end{array}
$$

Since

$$g \circ p_n \circ q_n = f \circ q_n$$

is null homotopic by the assumption, there exists a map

$$h\colon \Sigma(X|*)^n \longrightarrow Y$$

such that $g \simeq h \circ \partial$. From the suspension splitting, the inclusion

$$\Sigma(X|*)^n \hookrightarrow \Sigma X^n$$

admits a retraction $\Sigma X^n \to \Sigma(X|*)^n$. Thus the boundary map

$$\partial\colon X^{(n)} \longrightarrow \Sigma(X|*)^n$$

is null homotopic. It follows the map g is null homotopic and so is f. The induction is finished and hence the result. $\square$

COROLLARY 1.1.4. *Let X and Y be pointed spaces. Suppose that X is a suspension or Y is a loop space. Then the function*

$$q_n^*\colon [J_n(X), Y] \longrightarrow [X^n, Y]$$

is one-to-one.

PROOF. [1] If Y is a loop space, the assertion follows from Proposition 1.1.3 because $q_n^*\colon [J_n(X), Y] \to [X^n, Y]$ is a group homomorphism. If X is a suspension, the proof is given by induction on n. Clearly the statement holds for $n = 1$. Suppose that the statement holds for $n - 1$ with $n \geq 2$. Let $f, g\colon J_n(X) \to Y$ be any maps such that $f \circ q_n \simeq g \circ q_n$. Since

$$f|_{J_{n-1}(X)} \circ q_{n-1} = f \circ q_n|_{X^{n-1}} \simeq g \circ q_n|_{X^{n-1}} = g|_{J_{n-1}(X)} \circ q_{n-1},$$

$f|_{J_{n-1}(X)} \simeq g|_{J_{n-1}(X)}$ by induction. By Corollary 1.1.2, there is a map $h\colon X^{(n)} \to Y$ such that the homotopy class $[g] = [f] \cdot [h]$ in $[J_n(X), Y]$. It follows that

$$[f \circ q_n] = [g \circ q_n] = [f \circ q_n] \cdot [h]$$

in $[X^n, Y]$. It is routine to check that the action of $[X^{(n)}, Y]$ on $[X^n, Y]$ is free. Thus $[h] = 1$ and so $[f] = [g]$. The induction is finished and hence the result. $\square$

Let X and Y be pointed spaces. The map

$$d^i\colon X^n \to X^{n+1}, \quad (z_0, z_1, \ldots, z_{n-1}) \mapsto (z_0, \ldots, z_{i-1}, *, z_i, \ldots, z_{n-1})$$

induces a function (group homomorphism if Y is a loop space)

$$d_i = d^{i*}\colon [X^{n+1}, Y] \to [X^n, Y]$$

for $0 \leq i \leq n$. Recall that a sequence of sets (groups) $\mathcal{S} = \{S_n\}_{n \geq 0}$ is called a Δ-*set* if there are functions (group homomorphisms) $d_i\colon S_n \to S_{n-1}$, $0 \leq i \leq n$, such that the identity

(1.1.2) $$d_j d_i = d_i d_{j+1}$$

holds for all $j \geq i$. Let $\mathfrak{K}(X, Y)$ denote the sequence of sets $\{[X^{n+1}, Y]\}_{n \geq 0}$ with faces $d_i = d^{i*}$. It is straight forward to check that the above equality holds for d_i and so $\mathfrak{K}(X, Y)$ is a Δ-set (Δ-group if Y is a loop space).

Let $\mathcal{S}$ be a Δ-set. The *Cohen set* (*Cohen group* if $\mathcal{S}$ is a Δ-group) $\mathfrak{H}_n \mathcal{S}$ is defined by

$$\mathfrak{H}_n \mathcal{S} = \{x \in S_n \mid d_0 x = d_1 x = \cdots = d_n x\},$$

---

[1]The proof was omitted in the original manuscript. The author would like to thank referee's suggestion to include a proof to this corollary. This proof is also outlined by the referee.

namely, $\mathfrak{H}_n\mathcal{S}$ is the equalizer of the faces d_i for $0 \leq i \leq n$. Observe that the face $d_0\colon S_n \to S_{n-1}$ induces $p_n\colon \mathfrak{H}_n\mathcal{S} \to \mathfrak{H}_{n-1}\mathcal{S}$ such that

$$\begin{array}{ccc} S_n & \xrightarrow{d_i} & S_{n-1} \\ \uparrow & & \uparrow \\ \mathfrak{H}_n\mathcal{S} & \xrightarrow{p_n} & \mathfrak{H}_{n-1}\mathcal{S} \end{array}$$

commutes for each $0 \leq i \leq n$. The *total Cohen set* (*total Cohen group* if $\mathcal{S}$ is a Δ-group) $\mathfrak{H}\mathcal{S}$ is defined to be the inverse limit

$$\mathfrak{H}\mathcal{S} = \lim_{p_n} \mathfrak{H}_n\mathcal{S}.$$

For the Δ-set $\mathfrak{K}(X,Y)$, the quotient map $q_n\colon X^n \to J_n(X)$ induces functions

$$q_n^*\colon [J_n(X), Y] \longrightarrow \mathfrak{H}_{n-1}\mathfrak{K}(X,Y),$$

$$[J(X), Y] \longrightarrow \lim_n [J_n(X), Y] \xrightarrow{\lim_n q_n^*} \mathfrak{H}\mathfrak{K}(X,Y).$$

THEOREM 1.1.5. *Let X and Y be path-connected spaces. Suppose that X is a co-H-space or Y is an H-space. Then*

1) *the function $q_n^*\colon [J_n(X), Y] \longrightarrow \mathfrak{H}_{n-1}\mathfrak{K}(X,Y)$ is one-to-one and onto;*
2) *the function $[J(X), Y] \longrightarrow \mathfrak{H}\mathfrak{K}(X,Y)$ is one-to-one and onto.*

PROOF OF THEOREM 1.1.5 IN THE CASE THAT Y IS AN H-SPACE. If Y is a loop space, a proof of assertion (1) can be found in [**48**, Lemma 2.9] and then assertion (2) follows easily from the suspension splitting theorem of $J(X)$. Now assume that Y is a path-connected H-space. Then the inclusion $j\colon Y \to \Omega\Sigma Y$ admits a retraction $r\colon \Omega\Sigma Y \to Y$. The assertions follows from the commutative diagram

$$\begin{array}{ccc} [J_n(X), Y] & \longrightarrow & \mathfrak{H}_{n-1}\mathfrak{K}(X,Y) \\ \downarrow j_* & & \downarrow j_* \\ [J_n(X), \Omega\Sigma Y] & \xrightarrow{\cong} & \mathfrak{H}_{n-1}\mathfrak{K}(X, \Omega\Sigma Y) \\ \downarrow r_* & & \downarrow r_* \\ [J_n(X), Y] & \longrightarrow & \mathfrak{H}_{n-1}\mathfrak{K}(X,Y) \end{array}$$

with $r_* \circ j_* = \mathrm{id}$. □

We are going to prove the theorem in the case that X is a co-H-space. We need some lemmas.

For each sequence $I = (i_1, i_2, \ldots, i_k)$ with $0 \leq i_1 < \cdots < i_k \leq n-1$, let

$$d^I\colon X^k \to (X|_{n-k*})^n$$

denote the composite $d^{i_k} \circ \cdots \circ d^{i_1}$, where $d^i\colon X^s \to X^{s+1}$ is defined above. In other words, $d^I(x_1, x_2 \ldots x_k) \in (X|_{n-k*})^n$ is a point whose $(i_s + 1)^{\mathrm{st}}$ coordinate is

x_s, $1 \leq s \leq k$, and the rest coordinates are the basepoint. (For instance, for $k=1$ and $n=2$, $d^0(x_1) = (x_1, *)$ and $d^1(x_1) = (*, x_1)$.) Let I run over all sequences $(i_1, \ldots, i_k)$ with $0 \leq i_1 < \cdots < i_k \leq n-1$. The maps $d^I \colon X^k \to (X|_{n-k}*)^n$ define a map

$$\theta_k^n \colon \bigvee_{0 \leq i_1 < i_2 < \cdots < i_k \leq n-1} X^k \longrightarrow (X|_{n-k}*)^n.$$

The proof of the following lemma is similar to that of Proposition 1.1.3.

LEMMA 1.1.6. *Let X and Y be pointed spaces. Suppose that $n \geq k \geq 1$. Then the kernel of the function*

$$\theta_k^{n*} \colon [(X|_{n-k}*)^n, Y] \longrightarrow \prod_{0 \leq i_1 < i_2 < \cdots < i_k \leq n-1} [X^k, Y] = [\bigvee_{0 \leq i_1 < i_2 < \cdots < i_k \leq n-1} X^k, Y]$$

is trivial. □

From the commutative diagram of cofibre sequences

we have the following:

LEMMA 1.1.7. *Let X and Y be pointed spaces. Suppose that $n \geq k \geq 1$. Then the function*

$$\theta_k^{n*} \colon [(\Sigma X|_{n-k}*)^n, Y] \longrightarrow \prod_{0 \leq i_1 < i_2 < \cdots < i_k \leq n-1} [(\Sigma X)^k, Y]$$

is one-to-one. □

Let

$$s \colon J_n(X) \vee J_n(X) = J_n(X) \rtimes \partial I \hookrightarrow (J_{n-1}(X) \rtimes I) \cup (J_n(X) \rtimes \partial I)$$

be the inclusion of subspaces of $X \rtimes I$, where $I = [0,1]$ with $\partial I = \{0, 1\}$ and the half-smash $A \rtimes B = (A \times B)/(* \times B)$.

LEMMA 1.1.8. *Suppose that X is a suspension. Then*

$$s^* \colon [(J_{n-1}(X) \rtimes I) \cup (J_n(X) \rtimes \partial I), Y] \longrightarrow [J_n(X) \vee J_n(X), Y]$$

is one-to-one for any space Y.

PROOF. Let $X = \Sigma Z$ and let A denote $(J_{n-1}(X) \rtimes I) \cup (J_n(X) \rtimes \partial I)$. Consider the commutative diagram

$$\begin{array}{ccccc}
J_n(X) = J_n(X) \rtimes 0 & =\!=\!= & J_n(X) & \longrightarrow & * \\
\downarrow\scriptstyle{s_0} & & \downarrow\scriptstyle{\simeq} & & \downarrow \\
A & \xhookrightarrow{j} & J_n(X) \rtimes I & \longrightarrow\!\!\!\!\rightarrow & X^{(n)} \wedge S^1 \\
\uparrow\scriptstyle{\phi} & & \uparrow & & \| \\
((CZ|Z)^n \rtimes I) \cup ((CZ)^n \rtimes \partial I) & \hookrightarrow & (CZ)^n \rtimes I \simeq * & \longrightarrow\!\!\!\!\rightarrow & X^{(n)} \wedge S^1,
\end{array}$$

where the rows are cofibrations. The commutative diagram implies that the composite

$$J_n(X) \bigvee (((CZ|Z)^n \rtimes I) \cup ((CZ)^n \rtimes \partial I)) \xrightarrow{s_0 \vee \phi} A \vee A \xrightarrow{\text{fold}} A$$

is a homotopy equivalence. Now consider the homotopy commutative diagram

$$\begin{array}{ccccccc}
J_n(X) \rtimes 0 & \hookrightarrow & J_n(X) \rtimes \partial I & \longrightarrow & J_n(X) \rtimes 1 = J_n(X) & \xrightarrow{p_n} & X^{(n)} \\
\| & & \downarrow\scriptstyle{s_0 \cup s_1} & & \downarrow & & \| \\
J_n(X) \rtimes 0 & \xhookrightarrow{s_0} & A & \xrightarrow{r} & ((CZ|Z)^n \rtimes I) \cup ((CZ)^n \rtimes \partial I) & \xrightarrow{\simeq} & X^{(n)},
\end{array}$$

where p_n is the quotient map, r is a retraction of the map ϕ and the rows except the last terms are splitting cofibre sequences. Since the boundary map

$$X^{(n)} \longrightarrow \Sigma J_{n-1}(X)$$

is null homotopic, the function

$$p_n^* \colon [X^{(n)}, Y] \longrightarrow [J_n(X), Y]$$

is one-to-one and so, from the above commutative diagram, the function

$$(s_0 \cup s_1)^* \colon [A, Y] \longrightarrow [J_n(X) \rtimes \partial I, Y] = [J_n(X) \vee J_n(X), Y]$$

is one-to-one. This finishes the proof. $\square$

PROOF OF THEOREM 1.1.5 IN THE CASE THAT X IS A CO-H-SPACE. Since any path-connected co-H-space is a retract of a suspension, we may assume that X is a suspension.

(1). By Corollary 1.1.4, the function

$$q_n^* \colon [J_n(X), Y] \longrightarrow \mathfrak{H}_{n-1}\mathfrak{K}(X, Y)$$

is one-to-one. We show that q_n^* is onto by induction on n. Clearly q_1^* is onto. Suppose that q_k^* is onto for $k < n$ with $n \geq 2$. Let $f \colon X^n \to Y$ be a map such that the homotopy class $[f]$ lies in $\mathfrak{H}_{n-1}\mathfrak{K}(X, Y)$, that is,

$$f \circ d^0 \simeq f \circ d^1 \simeq \cdots \simeq f \circ d^{n-1} \colon X^{n-1} \longrightarrow Y.$$

Since $[f \circ d^0] = d_0[f] \in \mathfrak{H}_{n-2}\mathfrak{K}(X,Y)$, there is a map $g\colon J_{n-1}(X) \to Y$ such that
$$f \circ d^0 \simeq g \circ q_{n-1}$$
by induction. From the pushout diagram

$$\begin{CD}
(X|*)^n @>{j}>{\text{cofibration}}> X^n \\
@V{\tilde{q}_n}VV @VV{q_n}V \\
J_{n-1}(X) @>>> J_n(X),
\end{CD}$$

it suffices to show that
$$f \circ j \simeq g \circ \tilde{q}_n\colon (X|*)^n \longrightarrow Y$$
because, if so, the map f will factors through q_n up to homotopy. Let
$$\theta_{n-1}^n\colon \bigvee_{s=1}^n X^{n-1} \longrightarrow (X|*)^n$$
be the map defined in Lemma 1.1.6. Then, in the set $\left[\bigvee_{s=1}^n X^{n-1}, Y\right] = \prod_{s=1}^n [X^{n-1}, Y]$,
$$[g \circ \tilde{q}_n \circ \theta_{n-1}^n] = ([g \circ q_{n-1}], \ldots, [g \circ q_{n-1}]) = ([f \circ d^0], [f \circ d^1], \ldots, [f \circ d^{n-1}]) = [f \circ j \circ \theta_{n-1}^n]$$
because
$$d_i[f] = d_0[f]$$
for all i. By Lemma 1.1.7, θ_{n-1}^{n*} is one-to-one and so
$$g \circ \tilde{q}_n \simeq f \circ j.$$
Thus assertion (1) follows.

(2). It suffices to show that the function
$$[J(X), Y] \longrightarrow \lim_n [J_n(X), Y]$$
is one-to-one. Let $f, g\colon J(X) \to Y$ be maps such that
$$f|_{J_n(X)} \simeq g|_{J_n(X)}\colon J_n(X) \longrightarrow Y$$
for all n. We are going to construct homotopies $F_n\colon J_n(X) \rtimes I \to Y$ between $f|_{J_n(X)}$ and $g|_{J_n(X)}$ such that $F_{n+1}|_{J_n(X) \rtimes I} = F_n$ by induction on n starting with a homotopy $F_1: f|_X \simeq g|_X\colon X \rtimes I \to Y$. Suppose that $F_{n-1}\colon J_{n-1}(X) \rtimes I \to Y$ is defined such that $F_{n-1}(x, 0) = f(x)$ and $F_{n-1}(x, 1) = g(x)$ for $x \in J_{n-1}(X)$. Let
$$G = f \cup F_{n-1} \cup g\colon (J_n(X) \rtimes 0) \cup (J_{n-1}(X) \rtimes I) \cup (J_n(X) \rtimes 1) \longrightarrow Y$$
and let $F\colon J_n(X) \rtimes I \to Y$ be a homotopy between f and g. Observe that
$$G(x) = F(x) = (f \cup g)(x)$$
for $x \in J_n(X) \rtimes \partial I = J_n(X) \vee J_n(X)$. By Lemma 1.1.8, the map F restricted to the subspace $(J_n(X) \rtimes \partial I) \cup (J_{n-1}(X) \rtimes I)$ is homotopic to G. Since the inclusion
$$(J_n(X) \rtimes \partial I) \cup (J_{n-1}(X) \rtimes I) \hookrightarrow J_n(X) \rtimes I$$

is a cofibration, the map G admits an extension $F_n\colon J_n(X)\rtimes I\to Y$. The induction is completed and so the map

$$\tilde{F}=\bigcup_n F_n\colon J(X)\longrightarrow Y$$

is a well-defined homotopy between f and g. This finishes the proof of Theorem 1.1.5 □

The following universal property is useful.

PROPOSITION 1.1.9. *Let $f\colon X\to\Omega Y$ be a pointed map. Then there exists an unique, up to homotopy, H-map $Jf\colon J(X)\to\Omega Y$ such that $Jf|_X=f$.*

PROOF. Clearly any pointed map $f\colon X\to\Omega Y$ admits an extension of H-map from $J(X)\to\Omega Y$. It suffices to show the uniqueness. Let $\tilde{f},f'\colon J(X)\to\Omega Y$ be two H-maps such that $\tilde{f}|_X\simeq f'|_X\simeq f\colon X\to\Omega Y$. Consider the homomorphisms of monoids

$$\tilde{f}_*,f'_*\colon [X^n,J(X)]\longrightarrow [X^n,\Omega Y].$$

Let $x_i\in[X^n,J(X)]$ represented by the composite

$$X^n\xrightarrow{\pi_i}X\hookrightarrow J(X),$$

where π_i is the ith coordinate projection. Then

$$\tilde{f}_*(x_i)=f'_*(x_i)=[f\circ\pi_i]$$

for $1\leq i\leq n$. Thus

$$\tilde{f}_*(x_1x_2\cdots x_n)=f'_*(x_1x_2\cdots x_n)\in[X^n,\Omega Y]$$

for each n. Observe that the element $x_1x_2\cdots x_n$ is represented by the composite

$$X^n\xrightarrow{q_n}J_n(X)\hookrightarrow J(X)$$

by the canonical multiplication of $J(X)$. We obtain that

$$\tilde{f}|_{J_n(X)}\circ q_n\simeq f'|_{J_n(X)}\circ q_n$$

for each n. By Corollary 1.1.4,

$$\tilde{f}|_{J_n(X)}\simeq f'|_{J_n(X)}\colon J_n(X)\to\Omega Y$$

for each n. From Theorem 1.1.5, $\tilde{f}\simeq f'\colon J(X)\to\Omega Y$ and hence the result. □

1.2. Bi-Δ-groups

In this section, some properties of bi-Δ-groups are investigated. A sequence of sets $\mathcal{S}=\{S_n\}_{n\geq 0}$ is called a bi-Δ-set if there are faces $d_j\colon S_n\to S_{n-1}$ and cofaces $d^j\colon S_{n-1}\to S_n$ for $0\leq j\leq n$ such that the following identities hold:

(1.2.1) $\qquad\qquad d_jd_i=d_id_{j+1}\quad\text{for}\quad j\geq i,$

(1.2.2) $\qquad\qquad d^jd^i=d^{i+1}d^j\quad\text{for}\quad j\leq i,$

(1.2.3) $\qquad\qquad d_jd^i=\begin{cases}d^{i-1}d_j & \text{for}\quad j<i,\\ \text{id} & \text{for}\quad j=i,\\ d^id_{j-1} & \text{for}\quad j>i,\end{cases}$

where $d_0x=*$ for $x\in S_0$. In other words, $\mathcal{S}$ is a Δ- and co-Δ-set such that relation (1.2.3) holds. Moreover a sequence of groups $\mathcal{G}$ is called a *bi-Δ-group* if

$\mathcal{G}$ is a bi-Δ-set such that all faces and cofaces are group homomorphisms. A *weak bi-Δ-group* means a bi-Δ-set $\mathcal{G} = \{G_n\}_{n\geq 0}$ such that each G_n is a group and all faces are group homomorphisms, namely the cofaces are not required to be group homomorphisms.

Recall that the *Moore complex* $N\mathcal{G} = \{N_n G\}_{n\geq 0}$ of a Δ-group $\mathcal{G}$ is defined by

$$N_n \mathcal{G} = \bigcap_{i=1}^{n} \mathrm{Ker}(d_i \colon G_n \to G_{n-1})$$

with $d_0 \colon N_n \mathcal{G} \to N_{n-1}\mathcal{G}$. The groups of the *Moore cycles* $\mathcal{Z}\mathcal{G}$ and the *Moore boundaries* $\mathcal{B}_n \mathcal{G}$ are defined by

$$\mathcal{Z}_n \mathcal{G} = \bigcap_{i\geq 0} \mathrm{Ker}(d_i \colon G_n \to G_{n-1}) \text{ and } \mathcal{B}_n \mathcal{G} = d_0(N_{n+1} G),$$

respectively. The *Moore homotopy group* $\pi_n(\mathcal{G})$ is defined to be the coset

$$\pi_n(G) = H_n(N\mathcal{G}, d_0) = \mathcal{Z}_n \mathcal{G}/\mathcal{B}_n \mathcal{G}.$$

(**Note.** $\mathcal{B}_n \mathcal{G}$ need not be normal in $\mathcal{Z}_n \mathcal{G}$ for general Δ-group. Some properties of Δ-groups have been studied in [**3**].) The homotopy groups of bi-Δ-groups seem less interesting as one can easily check that the homotopy groups of a bi-Δ-group are all trivial. But the cycles in bi-Δ-groups are interesting, for instance, if $\mathcal{G}$ is the bi-Δ-group given by

$$G_n = [T^{n+1}, \Omega Y]$$

for $n \geq 0$, then

$$\mathcal{Z}_{n-1}\mathcal{G} = \pi_n(\Omega Y) = \pi_{n+1}(Y),$$

where T^n is the n-fold Cartesian product of S^1.

The Cohen group $\mathfrak{H}\mathcal{G}$ of a weak bi-Δ-group $\mathcal{G}$ is always a progroup by the following proposition.

PROPOSITION 1.2.1. *Let $\mathcal{G}$ be a weak bi-Δ-group. Then the map*

$$p_n \colon \mathfrak{H}_n \mathcal{G} \longrightarrow \mathfrak{H}_{n-1}\mathcal{G}$$

is an epimorphism with the kernel $\mathcal{Z}_n \mathcal{G}$ for each n.

PROOF. The proof follows from the lines in the proof of [**48**, Lemma 2.10]. □

It should be pointed out that if $\mathcal{G}$ is a simplicial group, the map

$$p_n \colon \mathfrak{H}_n \mathcal{G} \longrightarrow \mathfrak{H}_{n-1}\mathcal{G}$$

is not onto in general and its cokernel is related to the homotopy groups $\pi_*(\mathcal{G})$. As a comparison between simplicial groups and bi-Δ-groups, we give the following statement on simplicial groups which seems interesting independently.

THEOREM 1.2.2. *Let $\mathcal{G} = \{G_n\}_{n\geq 0}$ be a reduced simplicial group, that is, $G_0 = \{1\}$. Then*

1) *$p_{2k+1} \colon \mathfrak{H}_{2k+1}\mathcal{G} \to \mathfrak{H}_{2k}\mathcal{G}$ is an epimorphism for each $k \geq 0$.*
2) *The image of $p_{2k} \colon \mathfrak{H}_{2k}\mathcal{G} \to \mathfrak{H}_{2k-1}\mathcal{G}$ is a normal subgroup of $\mathfrak{H}_{2k-1}\mathcal{G}$ with cokernel isomorphic to $\pi_{2k-1}(\mathcal{G})$ for each $k \geq 1$.*

PROOF. We refer the terminology on simplicial sets to [**14, 24**]. Let $\Delta[n]$ denote the standard n-simplex with the only non-degenerate element $\sigma_n \in \Delta[n]_n$. Let $E[n]$ be the quotient simplicial set of $\Delta[n]$ by requiring

$$d_i \sigma_n \sim d_0 \sigma_n$$

for each $0 \leq i \leq n$. Let $q_n \colon \Delta[n] \to E[n]$ be the quotient simplicial map and let $\bar{\sigma}_n = q_n(\sigma_n) \in E[n]_n$. Then $d_i \bar{\sigma}_n = d_0 \bar{\sigma}_n$ for all $0 \leq i \leq n$. By simplicial relations,

$$d_{i_1} d_{i_2} \cdots d_{i_k} \bar{\sigma}_n = d_{j_1} d_{j_2} \cdots d_{j_k} \bar{\sigma}_n$$

for all sequences $(i_1, i_2, \ldots, i_k)$ and $(j_1, j_2, \ldots, j_k)$. In other words, $E[n]$ has one-cell in each dimension up to n. Let $f_{d_i \sigma_n} \colon \Delta[n-1] \to \Delta[n]$ be the simplicial map by sending σ_{n-1} to $d_i \sigma_n$. Then there is a unique simplicial injection $j_n \colon E[n-1] \to E[n]$ such that the diagram

$$\begin{array}{ccc} \Delta[n-1] & \xrightarrow{f_{d_i \sigma_n}} & \Delta[n] \\ \downarrow q_{n-1} & & \downarrow q_n \\ E[n-1] & \xhookrightarrow{j_n} & E[n] \end{array}$$

commutes for each $0 \leq i \leq n$. In other words, $E[n-1]$ can be regarded as the simplicial subset of $E[n]$ by taking cells up to dimension $n-1$ under the map j_n.

Observe that

$$H_*(E[n]; \mathbb{Z}) = H_*(C),$$

where C is a chain complex given by $C_i = 0$ for $i > n$ and $C_i = \mathbb{Z}$ for $0 \leq i \leq n$ with the faces

$$\partial_k = \sum_{j=0}^{k} (-1)^j d_j(E[n]) = \sum_{j=0}^{k} (-1)^j \colon C_k \to C_{k-1}$$

for $1 \leq k \leq n$. It follows that $E[2k]$ is contractible for each $k \geq 0$ and the pinch map

$$E[2k+1] \longrightarrow E[2k+1]/E[2k] = S^{2k+1}$$

is a homotopy equivalence after geometric realization. Let $\mathrm{Hom}(X, Y)$ denote the set of simplicial maps. Given any simplicial set $\mathcal{S} = \{S_n\}_{n \geq 0}$. Recall that, for each $x \in S_n$, there is a unique simplicial map $f_x \colon \Delta[n] \to \mathcal{S}$, called *representing map* of x, such that $f_x(\sigma_n) = x$. Furthermore, the function

$$f \colon S_n \longrightarrow \mathrm{Hom}(\Delta[n], S) \qquad x \mapsto f_x$$

is one-to-one and onto (group isomorphism if $\mathcal{S}$ is a simplicial group), see [**14**] for details. Observe that for $x \in \mathfrak{H}_n \mathcal{G}$, the representing map $f_x \colon \Delta[n] \to \mathcal{G}$ factors through the quotient $E[n]$. This gives the commutative diagram

$$\begin{array}{ccc} G_n & \xrightarrow[\cong]{f} & \mathrm{Hom}(\Delta[n], \mathcal{G}) \\ \cup\uparrow & & \uparrow q_n^* \\ \mathfrak{H}_n \mathcal{G} & \xrightarrow[\cong]{f} & \mathrm{Hom}(E[n], \mathcal{G}). \end{array}$$

Now we start to prove the assertions: (1). Let $x \in \mathfrak{H}_{2k}\mathcal{G}$ and let
$$f_x \colon E[2k] \to \mathcal{G}$$
be the representing map of x. Since $E[2k]$ is contractible and $\mathcal{G}$ is fibrant, there is an extension map $g \colon E[2k+1] \to \mathcal{G}$ such that $g|_{E[2k]} = f_x \colon E[2k] \to \mathcal{G}$. Let $y = g(\bar{\sigma}_{2k+1})$. Then
$$d_i y = g(d_i \bar{\sigma}_{2k+1}) = g(\bar{\sigma}_{2k}) = f_x(\bar{\sigma}_{2k}) = x$$
for $0 \leq i \leq 2k+1$. This proves that $\mathfrak{H}_{2k+1}\mathcal{G} \to \mathfrak{H}_{2k}\mathcal{G}$ is an epimorphism.

(2). Let $x \in \mathfrak{H}_{2k-1}\mathcal{G}$ and let $f_x \colon E[2k-1] \to \mathcal{G}$ be the representing map. Consider the cofibration
$$E[2k-1] \hookrightarrow E[2k].$$
Since $E[2k]$ is contractible, the map f_x extends to $E[2k]$ if and only if f_x is null homotopic. Thus
$$K = \mathrm{Im}(\mathfrak{H}_{2k}\mathcal{G} \longrightarrow \mathfrak{H}_{2k-1}\mathcal{G}) = \{x \in \mathfrak{H}_{2k-1}\mathcal{G} \mid f_x \simeq *\}.$$
Let $x, y \in \mathfrak{H}_{2k-1}\mathcal{G}$ be represented by f_x and g_x respectively. The the commutator $[x,y] = x^{-1}y^{-1}xy$ is represented by the composite
$$E[2k-1] \xrightarrow{\bar{\Delta}} E[2k-1] \wedge E[2k-1] \xrightarrow{f_x \wedge g_x} \mathcal{G} \wedge \mathcal{G} \xrightarrow{[-,-]} \mathcal{G},$$
since $E[2k-1] \simeq S^{2k-1}$ with $2k-1 \geq 1$, the reduced diagonal
$$\bar{\Delta} \colon E[2k-1] \to E[2k-1] \wedge E[2k-1]$$
is null homotopic. It follows that the commutator subgroup $[\mathfrak{H}_{2k-1}\mathcal{G}, \mathfrak{H}_{2k-1}\mathcal{G}]$ is contained in K. In particular, K is a normal subgroup of $\mathfrak{H}_{2k-1}\mathcal{G}$ with
$$\mathfrak{H}_{2k-1}\mathcal{G}/K \cong \mathrm{Hom}(E[2k-1], \mathcal{G})/\{f \colon E[2k-1] \to \mathcal{G} \mid f \simeq *\}$$
$$\cong [E[2k-1], \mathcal{G}] \cong \pi_{2k-1}(\mathcal{G}).$$
$\square$

REMARK 1.2.3. We gives some remarks to Theorem 1.2.2:
1) The *reduced* condition of $\mathcal{G}$ is used for technical reason that any simplicial map $f \colon X \to \mathcal{G}$ is automatically a pointed simplicial map and any homotopy $F \colon X \times I \to \mathcal{G}$ is automatically a pointed homotopy. Possibly this condition can be removed.
2) Let $E = \bigcup_{n=0}^{\infty} E[n]$. Then $\mathfrak{H}\mathcal{G} \cong \mathrm{Hom}(E, \mathcal{G})$. The group $\mathfrak{H}\mathcal{G}$ depends on simplicial group $\mathcal{G}$ rather than its homotopy type. But the cokernel of $\mathfrak{H}_{2k}\mathcal{G} \to \mathfrak{H}_{2k-1}\mathcal{G}$ only depends on the homotopy type of $\mathcal{G}$ as it is just the homotopy group.
3) This result somehow reveals that there is a possible way to kill off all even homotopy groups (or odd homotopy groups) by making certain constructions on simplicial groups. In our construction, $\mathfrak{H}\mathcal{G}$ is a tower of groups rather than the traditional simplicial groups.

Let $\mathcal{G}$ be a weak bi-Δ-group. For any integers $k \geq n$, the *James-Hopf operation* $H_{k,n} \colon G_k \to G_n$ is defined by $H_{k,k} = \mathrm{id}$ and, for $n > k$,
$$H_{k,n}(x) = \prod_{0 \leq i_1 < i_2 < \cdots < i_{n-k} \leq n} d^{i_{n-k}} d^{i_{n-k-1}} \cdots d^{i_1}(x)$$

with lexicographic order from right. If $x \in \mathcal{Z}_k\mathcal{G}$, then it is straightforward to check that
$$d_i H_{k,n}(x) = H_{k,n-1}(x)$$
for $0 \leq i \leq n$ and $k < n$, and so $H_{k,n}(x) \in \mathfrak{H}_n\mathcal{G}$ for any $n \geq k$ with
$$p_n H_{k,n}(x) = H_{k,n-1}(x).$$
This determines a unique element $H_k(x) \in \mathfrak{H}\mathcal{G}$ that projects to $H_{k,n}(x)$ in $\mathfrak{H}_n\mathcal{G}$ for each $n \geq k$ (in case $x \in \mathcal{Z}_k\mathcal{G}$).

Note. When $\mathcal{G} = \{[X^{n+1}, \Omega Y]\}_{n\geq 0}$, the subgroup $\mathcal{Z}_k\mathcal{G} \subseteq \mathfrak{H}_k\mathcal{G}$ is the group of the homotopy classes represented by the composites
$$J_{k+1}(X) \xrightarrow{\text{pinch}} X^{(k+1)} \xrightarrow{f} \Omega Y$$
for any map $f\colon X^{(k+1)} \to \Omega Y$. Given any element $x_f \in \mathcal{Z}_k\mathcal{G} \subseteq \mathfrak{H}_k\mathcal{G}$ represented by
$$x_f\colon J_{k+1}(X) \xrightarrow{\text{pinch}} X^{(k+1)} \xrightarrow{f} \Omega Y,$$
then the element $H_k(x_f) \in \mathfrak{H}\mathcal{G} = [J(X), \Omega Y]$ is represented by the composite
$$J(X) \xrightarrow{H_{k+1}} J(X^{(k+1)}) \xrightarrow{J(f)} \Omega Y,$$
where the map H_{k+1} is the James-Hopf invariants and $J(f)$ is the H-map such that $J(f)|_{X^{(k+1)}} = f$. If $\Omega Y = J(X^{(k+1)})$ and
$$f\colon X^{(k+1)} \to J(X^{(k+1)})$$
is the inclusion map, then $H_k(x_f)$ is the homotopy class $[H_{k+1}] \in [J(X), J(X^{(k+1)})]$. In general, the above composite can be rewritten as a formula
$$H_k(x_f) = Jf_*([H_{k+1}])$$
for any map $f\colon X^{(k+1)} \to \Omega Y$. This describes the operation H_k as the evaluations on the James-Hopf invariants in the case when $\mathcal{G} = \{[X^{n+1}, \Omega Y]\}_{n\geq 0}$.

The James-Hopf operation H_k has the following property.

THEOREM 1.2.4 (Taylor Series). *Let $\mathcal{G}$ be a weak bi-Δ-group. Then, for any $\alpha \in \mathfrak{H}\mathcal{G}$, there exists an unique element $\delta_k(\alpha) \in \mathcal{Z}_k G$ for $k \geq 0$ such that*
$$\alpha = \prod_{k=0}^{\infty} H_k(\delta_k(\alpha)).$$

PROOF. The functions
$$\phi_n\colon \prod_{k=0}^{n} \mathcal{Z}_k\mathcal{G} \xrightarrow{\prod_{k=0}^{n} H_{k,n}} \prod_{k=0}^{n} \mathfrak{H}_n\mathcal{G} \xrightarrow{\mu} \mathfrak{H}_n\mathcal{G}$$
induces a unique function
$$\phi = \lim_n \phi_n\colon \prod_{k=0}^{\infty} \mathcal{Z}_k\mathcal{G} \longrightarrow \mathfrak{H}\mathcal{G}.$$
It suffices to show by induction that ϕ_n is an isomorphism for each $n \geq 0$. Note that $\phi_0\colon \mathcal{Z}_0 G = G_0 \to G_0$ is the identity map. Suppose that ϕ_{n-1} is an isomorphism

with $n > 0$. From the commutative diagram of $\mathcal{Z}_n\mathcal{G}$-sets

$$\begin{array}{ccccc}
\mathcal{Z}_n\mathcal{G} & \hookrightarrow & \prod_{k=0}^{n} \mathcal{Z}_k\mathcal{G} & \twoheadrightarrow & \prod_{k=0}^{n-1} \mathcal{Z}_k\mathcal{G} \\
\| & & \downarrow \phi_n & \cong\downarrow \phi_{n-1} & \\
\mathcal{Z}_n\mathcal{G} & \hookrightarrow & \mathfrak{H}_n\mathcal{G} & \twoheadrightarrow & \mathfrak{H}_{n-1}\mathcal{G},
\end{array}$$

where $\mathcal{Z}_n\mathcal{G}$ acts on $\mathfrak{H}_n\mathcal{G}$ and $\prod_{k=0}^{n} \mathcal{Z}_k\mathcal{G}$ as subgroups and the short exactness of the bottom row is from Proposition 1.2.1, the function ϕ_n is an isomorphism and hence the result. $\square$

REMARK 1.2.5. We give some remarks to Theorem 1.2.4:

1) From the proof, the function $\delta_n\colon \mathfrak{H}\mathcal{G} \to \mathcal{Z}_k\mathcal{G}$ can be defined recursively as follows: $\delta_0(\alpha) = \alpha|_{\mathcal{G}_0} = \alpha|_{\mathfrak{H}_0\mathcal{G}}$, and

$$\delta_n(\alpha) = \alpha|_{\mathfrak{H}_n\mathcal{G}} \cdot H_{n-1,n}(\delta_{n-1}(\alpha))^{-1} \cdot H_{n-2,n}(\delta_{n-2}(\alpha))^{-1} \cdot \cdots \cdot H_{0,n}(\delta_0(\alpha))^{-1}.$$

The function δ_n provides a recursive way for constructing cycles from $\mathfrak{H}_n\mathcal{G}$. The simplicial analogues might be the algorithms for constructing higher dimensional cycles in simplicial groups such as Samelson products in [14].

2) Let $\phi\colon \mathcal{G} \to \mathcal{G}'$ be a morphism of weak bi-Δ-groups. Then

$$\phi(\alpha) = \prod_{k=0}^{\infty} H_k(\phi(\delta_k(\alpha)))$$

for any $\alpha \in \mathfrak{H}\mathcal{G}$. That is, the Taylor series is functorial.

3) Let $\mathcal{G} = \{[X^{n+1}, \Omega Y]\}_{n\geq 0}$ and let $\alpha \in \mathfrak{H}\mathcal{G} = [J(X), \Omega Y]$. Then

$$H_0(\delta_0(\alpha)) = J(\alpha)\colon J(X) \to \Omega Y$$

is the H-map induced by $\delta_0(\alpha) = \alpha|_X\colon X \to \Omega Y$. The higher terms

$$H_k(\delta_k(\alpha)) = J(\delta_k(\alpha)) \circ H_{k+1}$$

with $\delta_k(\alpha)$ being recursively given in (1). The Taylor series gives a decomposition formula of α and so it becomes an infinite summation in $[\Omega J(X), \Omega^2 Y]$ after looping.

4) Assume that X is path-connected. Let $\mathcal{G} = \{[X^{n+1}, \Omega\Sigma X]\}_{n\geq 0}$ and let

$$\alpha = q \in \mathfrak{H}\mathcal{G} = [J(X), \Omega\Sigma X]$$

be the qth power map of $J(X) \simeq \Omega\Sigma X$. Then the Taylor series gives a decomposition, called the *distributivity law* [8, Theorem 2.2.1],

$$q \simeq \Omega[q] \cdot \left(\prod_{k=1}^{\infty} J(\delta_k(q)) \circ H_{k+1}\right)$$

in $[\Omega\Sigma X, \Omega\Sigma X]$, where $[q]\colon \Sigma X \to \Sigma X$ is the map of degree q and

$$\delta_k(q)\colon X^{(k+1)} \to J(X)$$

is a product of certain $(k+1)$-fold iterated Samelson products. Suppose that $[p^r]\colon \Sigma X \to \Sigma X$ is null homotopic. Then

$$p^r = \prod_{k=1}^{\infty} H_k(\delta_k(p^r))$$

and so the Barratt conjecture is equivalent to that:

$$p \cdot \sum_{k=1}^{\infty} H_k(\delta_k(p^r))_* \colon \pi_*(\Omega\Sigma X) \longrightarrow \sum_{k=1}^{\infty} \pi_*(\Omega\Sigma X^{(k+1)}) \longrightarrow \pi_*(\Omega\Sigma X)$$

is a zero map if $[p^r]\colon \Sigma X \to \Sigma X$ is null homotopic. Thus the self maps

$$H_k(\delta_k(p^r))\colon \Omega\Sigma X \xrightarrow{H_{k+1}} \Omega\Sigma X^{(k+1)} \xrightarrow{J\delta_k(p^r)} \Omega\Sigma X$$

can be regarded as the obstructions to the Barratt conjecture.

5) Given an element $\alpha \in [J(X), \Omega Y]$, the explicit computation for the element

$$\delta_k(\alpha)\colon X^{(k+1)} \to \Omega Y$$

seems hard although it can be recursively defined by (1). However in the special cases when X is a sphere localized at 2, the 3-fold Whitehead products are trivial and so the Taylor series only has two terms as described in [**48**, Example 2.11]. Similarly if ΩY is a two-stage Postnikov system, the computations becomes relatively much easier which might help to understand the secondary operations on the cohomology of $J(X)$.

6) It should be also pointed out that, by using the functorial property of Taylor series, one may get information on the classical distributivity law by considering bi-Δ morphisms into (or from) the bi-Δ-group $\{[X^{n+1}, \Omega Y]\}_{n\geq 0}$. For instance, for spaces X with the property that $X = \Sigma X'$ and the degree map

$$[p^r]\colon X \to X$$

is null homotopic, the distributivity law for $[J(X), J(X)]$ lifts to the Cohen group introduced in [**11**]. The Cohen group for these spaces admits a faithful representation to the group of functorial self-coalgebra maps of tensor algebras over the ground ring $R = \mathbb{Z}/p^r$ according to Theorem 2.1.3. By considering convolution powers p^{r+t} of tensor algebras over the ground ring $R = \mathbb{Z}/p^r$, one obtains that the convolution power p^{r+t} is trivial restricted to the sub-coalgebra of tensor algebras having tensor-length $\leq p^{t+1} - 1$. It follows that the power map p^{r+t} of $J(X)$ is null homotopic restricted to $J_{p^{t+1}-1}(X)$. (See Example 2.1.8.)

LEMMA 1.2.6. *Let $\mathcal{G}$ be a weak bi-Δ-group. Then $\mathcal{G}$ satisfies the following strong fibrant condition*

Let $x_0, x_1, \cdots, x_n$ be elements in G_{n-1} such that $d_i x_j = d_{j-1} x_i$ for $i \geq j$. Then there exists an element $w \in G_n$ such that $d_i w = x_i$ for $0 \leq i \leq n$.

PROOF. We define the elements $w_i \in G_n$ for $0 \leq i \leq n$ recursively by putting

$$w_0 = d^0 x_0 \text{ and } w_i = w_{i-1}(d^i d_i w_{i-1})^{-1} d^i x_i$$

for $i > 0$. We show by induction that

$$d_j w_i = x_j$$

for $j \leq i$. The assertion will follow by taking $w = w_n$. Clearly $d_0 w_0 = y_0$. Suppose that
$$d_j w_{i-1} = x_j$$
for $j \leq i-1$ with $i > 0$. Then $d_i w_i = x_i$ and, for $j < i$,
$$d_j w_i = d_j w_{i-1} (d_j d^i d_i w_{i-1})^{-1} d_j d^i x_i = d_j w_{i-1} (d^{i-1} d_{i-1} d_j w_{i-1})^{-1} d^{i-1} d_j x_i$$
$$= x_j (d^{i-1} d_{i-1} x_j)^{-1} d^{i-1} d_{i-1} x_j = x_j.$$
The induction is finished and hence the result. $\square$

Let $\mathcal{G}$ be a Δ-group. Let $R_n \mathcal{G}$ be the sequence of groups defined by
$$(R_n \mathcal{G})_q = \begin{cases} 1 & 0 \leq q \leq n \\ \{x \in G_q \mid d_{i_1} d_{i_2} \cdots d_{i_{q-n}} x = 1 \text{ for all } (i_1, i_2, \ldots, i_{q-n})\} & q > n \end{cases}$$
with faces $d_i(R_n \mathcal{G}) = d_i(\mathcal{G})$. Observe that each $(R_n \mathcal{G})_q$ is a normal subgroup of G_q. Let $(P_n \mathcal{G})_q = G_q / (R_n \mathcal{G})_q$ for each $q \geq 0$. The sequence of Δ-groups

(1.2.4) $$\mathcal{G} \longrightarrow \cdots \longrightarrow P_n \mathcal{G} \longrightarrow P_{n-1} \mathcal{G} \longrightarrow \cdots \longrightarrow P_0 \mathcal{G}$$

is called the *Moore Postnikov system* of $\mathcal{G}$. Let $q_n \colon \mathcal{G} \to P_n \mathcal{G}$ be the quotient map. The following theorem tells that the progroup $\mathfrak{H}\mathcal{G}$ can be obtained by considering the Moore-Postnikov system.

THEOREM 1.2.7. *Let $\mathcal{G}$ be a weak bi-Δ-group. Then*
1) *each $R_n \mathcal{G}$ is a weak bi-Δ-subgroup and so the Moore-Postnikov system is a tower of weak bi-Δ-groups.*
2) $q_{n*} \colon \mathcal{Z}_k \mathcal{G} \to \mathcal{Z}_k P_n \mathcal{G}$ *is an isomorphism for $k \leq n$.*
3) $\mathcal{Z}_k P_n \mathcal{G} = 1$ *for $k > n$ and so $\mathfrak{H} P_n \mathcal{G} \cong \mathfrak{H}_n P_n \mathcal{G}$.*
4) $q_{n*} \colon \mathfrak{H}_n \mathcal{G} \to \mathfrak{H}_n P_n \mathcal{G}$ *is an isomorphism.*

PROOF. Assertion (1) is a routine exercise. Assertion (2) follows immediately from that $G_k = (P_n \mathcal{G})_k$ for $k \leq n$. Assertion (4) follows from assertions (2) and (3). We prove assertion (3). Let $x \in \mathcal{Z}_k P_n \mathcal{G}$ with $k > n$. Since $q_n \colon \mathcal{G} \to P_n \mathcal{G}$ is onto, there exists $y \in G_k$ such that $q_n(y) = x$. Since $d_j x = 1$ for all j, the elements
$$d_0 y, d_1 y, \cdots, d_k y$$
has matching faces in $(R_n \mathcal{G})_{k-1}$ and so, by Lemma 1.2.6, there exists an element $w \in (R_n \mathcal{G})_k$ such that
$$d_j w = d_j y$$
for $0 \leq j \leq k$. It follows that
$$y w^{-1} \in \mathcal{Z}_k G \subseteq (R_n \mathcal{G})_k$$
with $x = q_n(y w^{-1}) = 1$ and hence the result. $\square$

EXAMPLE 1.2.8. Let M be a manifold. Consider the configuration space
$$F(M, n+1) = \{(z_0, z_1, \ldots, z_n) \mid z_i \in M, \ z_i \neq z_j \text{ for } i \neq j\}.$$
Let $B(M, n+1) = F(M, n+1)/S_{n+1}$, where S_{n+1} acts on $F(M, n+1)$ by permuting coordinates. According to [**3**, Proposition 4.2.1], the sequence of fundamental groups
$$\mathcal{B}(M)^{\pi_1} = \{\pi_1(B(M, n+1))\}_{n \geq 0}$$

is a crossed Δ-group, that is, the face functions
$$d_i\colon \pi_1(B(M,n+1)) \longrightarrow \pi_1(B(M,n))$$
satisfies the rule that
$$d_i(\beta\beta') = d_i(\beta)d_{\beta(i)}(\beta'),$$
where the action of $\pi_1(B(M+1))$ on $\{0,1,\ldots,n\}$ is induced from the covering map $F(M,n+1) \to B(M,n+1)$. The i-th face d_i is essentially induced from the coordinate projection $F(M,n+1) \to F(M,n)$ by deleting $(i+1)$st coordinate. (See [**3**] for details.)

Suppose that M is a path-connected compact manifold with non-empty boundary ∂M. Let a be a point in ∂M. Then the map
$$F(M,n) \simeq F(M \smallsetminus a, n) \longrightarrow F(M, n+1),$$
$$(z_0, z_1, \ldots, z_{n-1}) \mapsto (z_0, \ldots, z_{i-1}, a, z_i, \ldots, z_{n-1}),$$
induces a group homomorphism
$$d^i\colon \pi_1(B(M,n)) \to \pi_1(B(M,n+1))$$
for $0 \leq i \leq n$. The sequence of groups $\mathcal{B}(M)^{\pi_1}$ with faces and cofaces is then a bi-Δ set with the property that $\mathcal{B}(M)^{\pi_1}$ is a co-Δ group under cofaces and a crossed Δ-group under faces. In particular, the sequence of pure braid groups
$$\{\pi_1(F(M,n+1))\}_{n \geq 0}$$
is a bi-Δ-group.

If $M = D^2$, then $\mathcal{B}(M)^{\pi_1} = \{B_{n+1}\}_{n \geq 0}$ is the sequence of the classical braid groups B_{n+1}. According to [**3**], the face operation d_i on the braids is obtained by deleting the $(i+1)$st string. From the above construction, the coface operation d^i is obtained by adding a trivial $(i+1)$st string in front of the other strings. Now let $\sigma_0, \sigma_1, \ldots, \sigma_{n-1}$ be the canonical generators for B_{n+1}. (**Note.** Our labelling system starts from 0 and so σ_i refers to σ_{i+1} in [**4**, p.8].) Then $d^i\colon B_n \to B_{n+1}$ is the group homomorphism such that
$$d^i \sigma_j = \begin{cases} \sigma_j & \text{for } j < i-1 \\ \sigma_i \sigma_{i-1} \sigma_i^{-1} & \text{for } j = i-1 \\ \sigma_{j+1} & \text{for } j > i-1. \end{cases}$$

The cycles $\mathcal{Z}_n \mathcal{B}(M)^{\pi_1}$ are the Brunnian braids on M, that is, the braids on M that become trivial after deleting any one of the strings. According to the graph given in [**4**, pp.70-71], the half-twist braid
$$\Delta_n = (\sigma_0 \sigma_1 \cdots \sigma_n)(\sigma_0 \sigma_1 \cdots \sigma_{n-1}) \cdots (\sigma_0 \sigma_1)(\sigma_0)$$
lies in $\mathfrak{H}_n \mathcal{B}(D^2)^{\pi_1}$ with the property that
$$d_i \Delta_n = \Delta_{n-1}$$
for $0 \leq i \leq n$. □

The following propositions are useful.

PROPOSITION 1.2.9 (Decomposition Theorem of bi-Δ-groups). *Let*
$$\mathcal{G} = \{G_n\}_{n \geq 0}$$
be a bi-Δ-group. Then G_n is the (iterated) semi-direct product the subgroups
$$d^{i_k} d^{i_{k-1}} \cdots d^{i_1}(\mathcal{Z}_{n-k}\mathcal{G}),$$

$0 \leq i_1 < \cdots < i_k \leq n$, $0 \leq k \leq n$, *with lexicographic from right.*

PROOF. The proof is given by induction on n. Clearly the statement holds for $n = 0$ because $\mathcal{Z}_0 \mathcal{G} = G_0$. Suppose that the statement holds for G_q with $0 \leq q < n$ for any bi-Δ-group $\mathcal{G}$. The exact sequence

$$1 \longrightarrow \mathrm{Ker}(d_n) \longrightarrow G_n \xrightarrow{d_n} G_{n-1}$$

admits a cross-section d^n and so

$$G_n = \mathrm{Ker}(d_n) \rtimes d^n(G_{n-1}).$$

Observe that the composite

$$G_0 \xrightarrow{d^{n-1}d^{n-2}\cdots d^0} G_n \xrightarrow{d_n} G_{n-1}$$

is trivial. There is a semi-direct decomposition

$$\mathrm{Ker}(d_n) = \mathrm{Ker}(d_n) \cap \mathrm{Ker}(d_0 d_1 \cdots d_{n-1}) \rtimes d^{n-1}d^{n-2}\cdots d^0(\mathcal{Z}_0 \mathcal{G}).$$

Let

$$(\Omega \mathcal{G})_{m-1} = \mathrm{Ker}(d_m \colon G_m \to G_{m-1}) \cap \mathrm{Ker}(d_0 d_1 \cdots d_{m-1})$$

with the induced faces and cofaces from $\mathcal{G}$. Then $\Omega \mathcal{G}$ is a bi-Δ-group with a semi-direct product decomposition

$$G_n \cong \mathrm{Ker}(d_n) \rtimes d^n(G_{n-1}) = (\Omega \mathcal{G})_{n-1} \rtimes d^{n-1}d^{n-2}\cdots d^0(\mathcal{Z}_0 \mathcal{G}) \rtimes d^n(G_{n-1}).$$

The assertion follows from the induction. □

Note. When $\mathcal{G} = \{\pi_1(F(M, n+1))\}_{n \geq 0}$ for a path-connected compact manifold M with nonempty boundary, then this lemma provides an algorithm for decomposing a pure braid on M as a product of Brunnian braids. If M is Riemann surface, then the group of Brunnian braids is free and so one get a different proof of the statement that the word problem for braid groups over Riemann surfaces is solvable by using this lemma.

PROPOSITION 1.2.10. *A sequence of bi-Δ-groups $\mathcal{G}' \to \mathcal{G} \to \mathcal{G}''$ is short exact if and only if the sequence of graded groups $\mathcal{Z}\mathcal{G}' \to \mathcal{Z}\mathcal{G} \to \mathcal{Z}\mathcal{G}''$ is short exact.*

PROOF. Suppose that

$$\mathcal{G}' \longrightarrow \mathcal{G} \xrightarrow{q} \mathcal{G}''$$

is short exact. It suffices to show that $\mathcal{Z}_n \mathcal{G} \to \mathcal{Z}_n \mathcal{G}''$ is onto for each $n \geq 0$. Let $x \in \mathcal{Z}_n \mathcal{G}''$ and let $\tilde{x}$ be a pre-image of x in $\mathcal{G}_n$. Then $d_i \tilde{x} \in \mathcal{G}'_{n-1}$ for $0 \leq i \leq n$. Since the elements $d_i \tilde{x}$ matching faces in $\mathcal{G}'$, there exists $w \in \mathcal{G}'_n$ such that $d_i w = d_i \tilde{x}$ for $0 \leq i \leq n$ by Lemma 1.2.6. Now the element $\tilde{x} w^{-1}$ lies in $\mathcal{Z}_n \mathcal{G}$ with $q(\tilde{x} w^{-1}) = x$ and hence $\mathcal{Z}_n \mathcal{G} \to \mathcal{Z}_n \mathcal{G}''$ is onto.

Now let

$$\mathcal{G}' \xrightarrow{j} \mathcal{G} \longrightarrow \mathcal{G}''$$

be a sequence of bi-Δ-groups such that

$$\mathcal{Z}\mathcal{G}' \longrightarrow \mathcal{Z}\mathcal{G} \longrightarrow \mathcal{Z}\mathcal{G}''$$

is exact. From the sequence of bi-Δ-groups

$$\mathrm{Ker}(j) \hookrightarrow \mathcal{G}' \twoheadrightarrow \mathrm{Im}(j) \hookrightarrow \mathcal{G} \twoheadrightarrow \mathrm{Im}(q) \hookrightarrow \mathcal{G}'',$$

there is a sequence of bi-Δ-groups

$$\mathcal{Z}\mathrm{Ker}(j) \hookrightarrow \mathcal{Z}\mathcal{G}' \twoheadrightarrow \mathcal{Z}\mathrm{Im}(j) \hookrightarrow \mathcal{Z}\mathcal{G} \twoheadrightarrow \mathcal{Z}\mathrm{Im}(q) \hookrightarrow \mathcal{Z}\mathcal{G}''.$$

Since $\mathcal{Z}\mathcal{G}' \to \mathcal{Z}\mathcal{G} \to \mathcal{Z}\mathcal{G}''$ is short exact, by using Proposition 1.2.9, $\mathrm{Ker}(j) = 1$, $\mathrm{Im}(j) = \mathrm{Ker}(q)$ and $\mathrm{Im}(q) = \mathcal{G}''$. The assertion follows. □

1.3. Skeletons of Bi-Δ-groups

In this section, we are going to investigate the skeleton filtration of bi-Δ-group. The construction given here is the categorical interpretation of the usual skeletons of CW-complexes or simplicial sets. Roughly speaking, the ideas for obtaining the skeletons are first to take the groups from a given bi-Δ-group up to dimension n with induced faces and cofaces, and then to blow up to be a bi-Δ-group in certain universal way.

An *n-partial bi-Δ-group* means a finite sequence of groups $\mathcal{G} = \{G_k\}_{0 \leq k \leq n}$ with faces and cofaces homomorphisms such that identities (1.2.1), (1.2.2) and (1.2.3) hold up to dimension n. Given an n-partial bi-Δ-group $\mathcal{G} = \{G_k\}_{0 \leq k \leq n}$, construct an $(n+1)$-partial bi-Δ-group $\Phi_n^{n+1}\mathcal{G}$ as follows:

Set the groups $(\Phi_n^{n+1}\mathcal{G})_k = G_k$ with the faces $d_i \colon (\Phi_n^{n+1}\mathcal{G})_k \to (\Phi_n^{n+1}\mathcal{G})_{k-1}$ and cofaces $d^i \colon (\Phi_n^{n+1}\mathcal{G})_{k-1} \to (\Phi_n^{n+1}\mathcal{G})_k$ induced from $\mathcal{G}$ for $0 \leq k \leq n$. Let $\Gamma^j = G_n$ be a copy of G_n for $0 \leq j \leq n+1$ and let

$$\Gamma = \coprod_{j=0}^{n+1} \Gamma^j$$

be the $(n+1)$-fold self free product of G_n. For $0 \leq i \leq n+1$, write $\tilde{d}^i \colon G_n \to \Gamma$ for the inclusion

$$G_n = \Gamma^i \hookrightarrow \Gamma = \coprod_{j=0}^{n+1} \Gamma^j.$$

Now define $(\Phi_n^{n+1}\mathcal{G})_{n+1}$ to be the quotient group of Γ modulo the following relations

(1.3.1) $$\tilde{d}^j \circ d^i(x) \equiv \tilde{d}^{i+1} d^j(x)$$

for all $0 \leq j \leq i \leq n$ and all $x \in G_{n-1}$, where $G_{-1} = \{1\}$ if $n = 0$. Let

$$q \colon \Gamma \to (\Phi_n^{n+1}\mathcal{G})_{n+1}$$

be the quotient map. The cofaces $d^i \colon G_n \to (\Phi_n^{n+1}\mathcal{G})_{n+1}$, $0 \leq i \leq n+1$, is defined to be the composite

$$d^i \colon G_n \xrightarrow{\tilde{d}^i} \Gamma \xrightarrow{q} (\Phi_n^{n+1}\mathcal{G})_{n+1}.$$

From the relations in the definition of $(\Phi_n^{n+1}\mathcal{G})_{n+1}$, identity (1.2.2) holds for $\Phi_n^{n+1}\mathcal{G}$ up to dimension $n+1$. The faces $d_i \colon (\Phi_n^{n+1}\mathcal{G})_{n+1} \to G_n$ will be defined such that $\Phi_n^{n+1}\mathcal{G}$ is an $(n+1)$-partial bi-Δ-group. Given any $0 \leq j \leq n+1$, since $\Gamma = \coprod\limits_{i=0}^{n+1} \Gamma^i$ is the free product, there is a unique group homomorphism $\tilde{d}_j \colon \Gamma \to G_n$ such that the composite

$$\tilde{d}_j \tilde{d}^i \colon G_n = \Gamma^i \hookrightarrow \Gamma \longrightarrow G_n$$

is given by the formula

(1.3.2) $$\tilde{d}_j \tilde{d}^i = \begin{cases} d^{i-1} d_j & i > j \\ \mathrm{id} & i = j \\ d^i d_{j-1} & i < j \end{cases}$$

for $0 \leq i \leq n+1$.

1.3. SKELETONS OF BI-Δ-GROUPS

LEMMA 1.3.1. *The group homomorphism $\tilde{d}_k \colon \Gamma \to G_n$ factors through the quotient $(\Phi_n^{n+1}\mathcal{G})_{n+1}$ for each $0 \leq k \leq n+1$.*

PROOF. The proof is a routine exercise. □

Let $d_i \colon (\Phi_n^{n+1}\mathcal{G})_{n+1} \to G_n$ be the homomorphism such that $\tilde{d}_i = d_i \circ q \colon \Gamma \to G_n$.

PROPOSITION 1.3.2. *Let $\mathcal{G}$ be an n-partial bi-Δ-group. Then $\Phi_n^{n+1}\mathcal{G}$ is an $(n+1)$-partial bi-Δ-group under the faces and cofaces defined as above.*

PROOF. Identities (1.2.1) and (1.2.3) follow from equations (1.3.2) and (1.3.1), respectively. Now show that identity (1.2.2) holds in $\Phi_n^{n+1}\mathcal{G}$, that is,
$$d_j d_i = d_i d_{j+1} \colon (\Phi_n^{n+1}\mathcal{G})_{n+1} \to G_{n-1}$$
for $i \leq j$. Since $q \colon \Gamma \to (\Phi_n^{n+1}\mathcal{G})_{n+1}$ is an epimorphism, it suffices to show that
$$d_j \tilde{d}_i = d_i \tilde{d}_{j+1} \colon \Gamma \to G_{n-1}$$
for $i \leq j$. Since $\Gamma = \coprod_{j=0}^{n+1} \Gamma^j$ is the free product, it suffices to show that, for $i \leq j$,
$$d_j \tilde{d}_i \tilde{d}^k = d_i \tilde{d}_{j+1} \tilde{d}^k \colon G_n \to G_{n-1}$$
for $0 \leq k \leq n+1$. Now it is a routine exercise to check the following formula:
$$d_j \tilde{d}_i \tilde{d}^k = d_i \tilde{d}_{j+1} \tilde{d}^k = \begin{cases} d^k d_{j-1} d_{i-1} = d^k d_{i-1} d_j & k < i \leq j \\ d_j & k = i \leq j \\ d^{k-1} d_{j-1} d_i = d^{k-1} d_i d_j & i < k \leq j \\ d_i & i \leq j = k-1 \\ d^{k-2} d_j d_i = d^{k-2} d_i d_{j+1} & i \leq j < k-1 \end{cases}$$
and hence the result. □

Let $\mathcal{G} = \{G_k\}_{0 \leq k < n+1}$ be an n-partial bi-Δ-group and let $\mathcal{G}' = \{G'_k\}_{0 \leq k < m+1}$ be an m-partial bi-Δ-group with $n \leq m \leq \infty$. (A bi-Δ-group is regarded as an ∞-partial bi-Δ-group.) A *pseudomorphism*
$$f = \{f_k\}_{0 \leq k < n+1} \colon \mathcal{G} \to \mathcal{G}'$$
means a sequence of group homomorphisms $f_k \colon G_k \to G'_k$, $0 \leq k < n+1$, such that $d_i f_k = f_{k-1} d_i$ for $0 \leq i \leq k < n+1$ and $d^i f_k = f_{k+1} d^i$ for $0 \leq i \leq k+1 < n+1$. (**Note.** If $m = n$, then a pseudomorphism is the same as a morphism of n-partial bi-Δ-groups.) The canonical inclusion $j_\mathcal{G} = \{\mathrm{id}_{G_k}\}_{0 \leq k < n+1} \colon \mathcal{G} \to \Phi_n^{n+1}\mathcal{G}$ is a pseudomorphism with the following universal property:

LEMMA 1.3.3. *Let $\mathcal{G}$ be an n-partial bi-Δ-group and let $\mathcal{G}'$ be an $(n+1)$-partial bi-Δ-group. Then for any pseudomorphism $f \colon \mathcal{G} \to \mathcal{G}'$ there exists a unique morphism of $(n+1)$-partial bi-Δ-groups $\tilde{f} \colon \Phi_n^{n+1}\mathcal{G} \to \mathcal{G}'$ such that $f = \tilde{f} \circ j_\mathcal{G}$.*

PROOF. The notations in the construction of $\Phi_n^{n+1}\mathcal{G}$ are used here. Since
$$\Gamma = \coprod_{j=0}^{n+1} \Gamma^j$$
is the free product, there exists a unique group homomorphism $g \colon \Gamma \to G'_{n+1}$ such that
$$g\tilde{d}^i = d^i f_n \colon G_n \to G'_{n+1}$$

for $0 \leq i \leq n+1$. Observe that, for $j \leq i$,
$$g\tilde{d}^j d^i = d^j f_n d^i = d^j d^i f_{n-1} = d^{i+1} d^j f_{n-1} = g\tilde{d}^{i+1} d^j : G_{n-1} \to G'_{n+1}.$$
The homomorphism $g \colon \Gamma \to G'_{n+1}$ factors through the quotient group $(\Phi_n^{n+1}\mathcal{G})_{n+1}$. Let $\tilde{f}_{n+1} \colon (\Phi_n^{n+1}\mathcal{G})_{n+1} \to G'_{n+1}$ be the resulting homomorphism. From the uniqueness of g, $\tilde{f}_{n+1}$ is the unique homomorphism with the property that
$$\tilde{f}_{n+1} d^i = d^i f_n$$
for $0 \leq i \leq n+1$. For $k > i$, we have
$$d_i g \tilde{d}^k = d_i d^k f_n = d^{k-1} d_i f_n = d^{k-1} f_{n-1} d_i = f_n d^{k-1} d_i = f_n d_i d^k = f_n d_i q \tilde{d}^k.$$
Similarly, the equality
$$d_i g \tilde{d}^k = f_n d_i q \tilde{d}^k$$
also holds for $k \geq i$. It follows that
$$d_i \tilde{f}_{n+1} q = d_i g = f_n d_i q \colon \Gamma \to G_n$$
for $0 \leq i \leq n+1$ because Γ is the free product. Thus
$$d_i \tilde{f}_{n+1} = f_n d_i$$
for $0 \leq i \leq n+1$ and hence $\tilde{f} \colon \Phi_n^{n+1}\mathcal{G} \to \mathcal{G}'$ is a morphism of $(n+1)$-partial bi-Δ-groups. This finishes the proof. □

Let $\mathcal{G}$ be an n-partial bi-Δ-group with $0 \leq n < \infty$. Let $\Phi_n^n \mathcal{G}$ denote $\mathcal{G}$ and, for $k \geq 1$, let $\Phi_n^{n+k}\mathcal{G} = \Phi_{n+k-1}^{n+k} \Phi_{n+k-2}^{n+k-1} \cdots \Phi_n^{n+1}\mathcal{G}$. Then $\Phi_n^{n+k}\mathcal{G}$ is an $(n+k)$-partial bi-Δ-group with the tower of pseudomorphisms
$$\mathcal{G} = \Phi_n^n \mathcal{G} \subseteq \Phi_n^{n+1}\mathcal{G} \subseteq \Phi_n^{n+2}\mathcal{G} \subseteq \cdots.$$
Let $\Phi_n \mathcal{G} = \bigcup_{k=0}^{\infty} \Phi_n^{n+k}\mathcal{G}$ be the colimit of the tower. Then $\Phi_n \mathcal{G}$ is a bi-Δ-group. By Lemma 1.3.3, the construction $\Phi_n \mathcal{G}$ has the following universal property:

PROPOSITION 1.3.4. *Let $\mathcal{G}$ be an n-partial bi-Δ-group with $0 \leq n < \infty$ and let $\mathcal{G}'$ be a bi-Δ-group. Then, for any pseudomorphism $f \colon \mathcal{G} \to \mathcal{G}'$, there exists a unique morphism of bi-Δ-groups $\tilde{f} \colon \Phi_n \mathcal{G} \to \mathcal{G}'$ such that $\tilde{f}|_{\mathcal{G}} = f$.* □

For any bi-Δ-group $\mathcal{G} = \{G_k\}_{k \geq 0}$, the n-partial bi-Δ-group $\mathrm{Par}_n \mathcal{G}$ is defined to be $\mathrm{Par}_n \mathcal{G} = \{G_k\}_{0 \leq k \leq n}$ with induced faces and cofaces from $\mathcal{G}$. The *n-skeleton* $\mathrm{sk}_n \mathcal{G}$ of $\mathcal{G}$ is defined to be
$$\mathrm{sk}_n \mathcal{G} = \Phi_n \mathrm{Par}_n \mathcal{G}$$
for each $n \geq 0$. By Proposition 1.3.4, for $m \leq n$, the inclusion $\mathrm{Par}_m \mathcal{G} \hookrightarrow \mathrm{Par}_n \mathcal{G}$ induces a unique morphism of bi-Δ-groups
$$\theta_m^n \colon \mathrm{sk}_m \mathcal{G} \to \mathrm{sk}_n \mathcal{G}$$
with the property that
$$\theta_n^q \circ \theta_m^n = \theta_m^q$$
for $m \leq n \leq q$.

THEOREM 1.3.5. *For each $0 \leq n \leq \infty$, there exists a functor sk_n from the category of bi-Δ-groups to itself together with natural transformations $\theta_m^n \colon \mathrm{sk}_m \to \mathrm{sk}_n$ such that the following holds:*

1) sk_∞ *is the identity functor.*

2) $\theta_n^q \theta_m^n = \theta_m^q$ for $0 \leq m \leq n \leq q \leq \infty$.
3) $\theta_n^\infty \colon (\mathrm{sk}_n \mathcal{G})_k \to (\mathrm{sk}_\infty \mathcal{G})_k = (\mathcal{G})_k$ is an isomorphism for $0 \leq k \leq n$ for any bi-Δ-group $\mathcal{G}$.
4) $\mathcal{G} = \mathrm{colim}_{\theta_n^{n+1}} \mathrm{sk}_n \mathcal{G}$ for any bi-Δ-group $\mathcal{G}$.
5) Let $f \colon \mathcal{G} \to \mathcal{G}'$ be a morphism of bi-Δ-groups. If $f_k \colon (\mathcal{G})_k \to (\mathcal{G}')_k$ is an epimorphism (isomorphism) for each $0 \leq k \leq n$, then
$$sk_n(f) \colon \mathrm{sk}_n \mathcal{G} \to \mathrm{sk}_n \mathcal{G}'$$
is an epimorphism (isomorphism).
6) $\mathrm{sk}_m \circ \mathrm{sk}_n = \mathrm{sk}_{\min\{m,n\}}$ for $m, n \geq 0$.

PROOF. Assertions (1)-(4) follows from the universal property of Φ_n. Assertion (5) follows from the definition of Φ_n. We prove assertion (6). Since
$$\mathrm{Par}_m \mathrm{sk}_n \mathcal{G} = \mathrm{Par}_m \mathcal{G}$$
for $m \leq n$, we have $\mathrm{sk}_m \circ \mathrm{sk}_n = \mathrm{sk}_m$ for $m \leq n$. Since $\mathrm{Par}_{n+1} \mathrm{sk}_n \mathcal{G} = \Phi_n^{n+1} \mathrm{Par}_n \mathcal{G}$,
$$\mathrm{sk}_n \mathcal{G} = \Phi_n^\infty \mathcal{G} = \Phi_{n+1}^\infty \Phi_n^{n+1} \mathcal{G} = \Phi_{n+1}^\infty \mathrm{Par}_{n+1} \mathrm{sk}_n \mathcal{G} = \mathrm{sk}_{n+1} \mathrm{sk}_n \mathcal{G}$$
for $n \geq 0$. By induction, $\mathrm{sk}_m \mathrm{sk}_n \mathcal{G} = \mathrm{sk}_n \mathcal{G}$ for all $m \geq n$ and hence the result. $\square$

Let $I(i_1, i_2, \ldots, i_k)$ be a sequence of nonnegative integers. We write d^I for the iterated cofaces $d^{i_1} d^{i_2} \cdots d^{i_k}$ and d_I for the iterated faces $d_{i_1} d_{i_2} \cdots d_{i_k}$. Note that $\theta_n^\infty \colon (\mathrm{sk}_n \mathcal{G})_k \to (\mathcal{G})_k$ is the identity for $k \leq n$. For $k > n$, the image of θ_n^∞ is described as follows:

PROPOSITION 1.3.6. Let $\mathcal{G} = \{G_m\}_{m \geq 0}$ be a bi-Δ-group and let $k > 0$. Then the image of
$$\theta_n^\infty \colon (\mathrm{sk}_n \mathcal{G})_{n+k} \to G_{n+k}$$
is the subgroup of G_{n+k} generated by the sets $d^I(G_n)$, where I runs over all sequences $I = (i_1, i_2, \ldots, i_k)$ with $n + k \geq i_1 > i_2 > \cdots > i_k \geq 0$.

PROOF. Recall that $\Phi_n^{n+1} \mathrm{Par}_n \mathcal{G}$ is a quotient group of the free product
$$\Gamma = \coprod_{j=0}^{n+1} \tilde{d}^j(G_n),$$
where the group $\tilde{d}^j(G_n) = \Gamma^j = G_n$. Since the map $(\Phi_n^{n+1} \mathrm{Par}_n \mathcal{G})_{n+1} \to G_{n+1}$ sends each Γ^j to $d^j(G_n)$, the image of
$$\theta_n^\infty \colon (\mathrm{sk}_n \mathcal{G})_{n+1} \to G_{n+1}$$
is the subgroup generated by $d^j(G_n)$ for $0 \leq j \leq n+1$. Suppose that the statement hold for less than k with $k > 1$. By iterating the construction Φ_s^{s+1}, the group
$$(\Phi_n^{n+k} \mathrm{Par}_n \mathcal{G})_{n+k} = (\Phi_{n+k-1}^{n+k} \Phi_{n+k-2}^{n+k-1} \cdots \Phi_n^{n+1} \mathrm{Par}_n \mathcal{G})_{n+k}$$
is a quotient group of the free product
$$\Gamma' = \coprod_{0 \leq i_k < i_{k-1} < \cdots < i_2 < i_1 \leq n+k} \tilde{d}^I(G_n),$$
where $\tilde{d}^I(G_n)$ is a copy of G_n indexed by iterated coface d^I. Since $\theta_n^\infty \colon \mathrm{sk}_n \mathcal{G} \to \mathcal{G}$ is a morphism of bi-Δ-groups, θ_n^∞ sends each $\tilde{d}^I(G_n)$ to the subgroup $d^I(G_n) \subseteq G_{n+k}$ and hence the result. $\square$

An explicit construction of the 0-skeleton $\mathrm{sk}_0\, \mathcal{G}$ is given. For any given group G, formally construct a bi-Δ-group $\mathfrak{F}^G$ as follows: The group

$$(\mathfrak{F}^G)_n = \coprod_{j=0}^{n} (G)_{x_j},$$

is the free product, where $(G)_{x_j}$ is a copy of G labelled by letters x_j. The face $d_i\colon (\mathfrak{F}^G)_n \to (\mathfrak{F}^G)_{n-1}$ is the (unique) group homomorphism such that $d_i|_{(G)_{x_i}}$ is the trivial map, $d_i|_{(G)_{x_j}}$ is the inclusion

$$(G)_{x_j} \hookrightarrow (\mathfrak{F}^G)_{n-1} = \coprod_{k=0}^{n-1} (G)_{x_k} \quad \text{if} \quad j < i$$

$$(G)_{x_j} = G = (G)_{x_{j-1}} \hookrightarrow (\mathfrak{F}^G)_{n-1} = \coprod_{k=0}^{n-1} (G)_{x_k} \quad \text{if} \quad j > i,$$

that is, d_i is the projection map that sends $(G)_{x_i}$ to the trivial groups and sends other copies of G down to $(\mathfrak{F}^G)_{n-1}$ in order. The coface $d^i\colon (\mathfrak{F}^G)_n \to (\mathfrak{F}^G)_{n+1}$ is the (unique) group homomorphism such that $d^i|_{(G)_{x_j}}$ is given by the inclusion.

$$(G)_{x_j} \hookrightarrow (\mathfrak{F}^G)_{n+1} = \coprod_{k=0}^{n+1} (G)_{x_k} \quad \text{if} \quad j < i$$

$$(G)_{x_j} = G = (G)_{x_{j+1}} \hookrightarrow (\mathfrak{F}^G)_{n+1} = \coprod_{k=0}^{n+1} (G)_{x_k} \quad \text{if} \quad j \geq i.$$

In other words, d^i is the ordered inclusion missing the copy $(G)_{x_i}$. It is straightforward to check that $\mathfrak{F}^G = \{(\mathfrak{F}^G)_n\}_{n\geq 0}$ is a bi-Δ-group under the faces and cofaces defined as above. (**Note.** The subgroup $(G)_{x_j}$ of $(\mathfrak{F}^G)_n$ is the image of the iterated coface

$$d^n d^{n-1} \cdots d^{j+1} d^{j-1} \cdots d^1 d^0 \colon (\mathfrak{F}^G)_0 = G \longrightarrow (\mathfrak{F}^G)_n$$

for $0 \leq j \leq n$.)

PROPOSITION 1.3.7. *The bi-Δ-group $\mathfrak{F}^G$ satisfies the following universal property:*

For any bi-Δ-group $\mathcal{G}'$ with a group homomorphism $f\colon G \to (\mathcal{G}')_0$, there is a unique morphism of bi-Δ-group $\tilde{f}\colon \mathfrak{F}^G \to \mathcal{G}'$ such that $\tilde{f}|_{(\mathfrak{F}^G)_0} = f$.

PROOF. The proof is a routine exercise, where

$$\tilde{f}\colon (\mathfrak{F}^G)_n = \coprod_{j=0}^{n} (G)_{x_j} \longrightarrow (\mathcal{G}')_n$$

is the unique group homomorphism such that $\tilde{f}|_{(G)_{x_j}}$ is given by the composite

$$G \xrightarrow{f} (\mathcal{G}')_0 \xrightarrow{d^n d^{n-1} \cdots d^{j+1} d^{j-1} \cdots d^1 d^0} (\mathcal{G}')_n,$$

namely, there is a unique morphism $\tilde{f}$ commuting with cofaces and then it is easy to check that $\tilde{f}$ also commutes with faces. □

By Proposition 1.3.4, we have the following:

COROLLARY 1.3.8. *There is an isomorphism of bi-Δ-groups*
$$\mathrm{sk}_0 \mathcal{G} \cong \mathfrak{F}^{(\mathcal{G})_0}$$
for any bi-Δ-group $\mathcal{G}$. □

In the particular case when $G = \mathbb{Z}$, the bi-Δ-group $\mathfrak{F}^\mathbb{Z}$ is a sequence of free groups with the following universal property:

COROLLARY 1.3.9. *Let $\mathcal{G}$ be any bi-Δ-group and let α be any element in $(\mathcal{G})_0$. Then there is a (unique) morphism of bi-Δ-groups*
$$e_\alpha \colon \mathfrak{F}^\mathbb{Z} \longrightarrow \mathcal{G}$$
such that $e_X \colon (\mathfrak{F}^\mathbb{Z})_0 = \mathbb{Z} \to (\mathcal{G})_0$ maps the generator of $\mathbb{Z}$ to α. Moreover if $\phi \colon \mathcal{G} \to \mathcal{G}'$ is a morphism of bi-Δ-groups, then the diagram

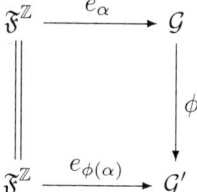

commutes for any $\alpha \in (\mathcal{G})_0$. □

Remarks. We give some remarks to the bi-Δ-group $\mathfrak{F}^\mathbb{Z}$.

1) As a group $(\mathfrak{F}^\mathbb{Z})_n$ is a free group of rank $n+1$ and faces are just given by the projections. Let $F(\Delta[1])$ be the Milnor $F(K)$-construction on the 1-simplex $\Delta[1]$. Recall from [47] that the simplicial group $F(\Delta[1])$ has same rank in dimension n as $\mathfrak{F}^\mathbb{Z}$ with coinciding faces. Thus, as Δ-groups, $\mathfrak{F}^\mathbb{Z} = F(\Delta[1])$.

2) The cycles $\mathcal{Z}F(\Delta[1]) \cong NF(S^1)$ have been determined in [46]. It was proved in [3] that the Moore chains $NF(S^1)$ are isomorphic to the group of Brunnian braids. A Brunnian braid means a braid which becomes trivial by deleting any one of its strings. For instance, the classical Borromean ring is a link obtained from a 3-string Brunnian braid.

3) From Proposition 1.2.1, $\mathfrak{H}\mathfrak{F}^\mathbb{Z} = \mathfrak{H}F(\Delta[1])$ is the progroup built by the cycles $\mathcal{Z}F(\Delta[1]) \cong NF(S^1)$. By interpreting into the language of braids [13], $\mathfrak{H}\mathfrak{F}^\mathbb{Z}$ is the progroup built by Brunnian braids. Since $\mathfrak{H}\mathfrak{F}^\mathbb{Z}$ can be also described as the group of functorial self maps of loop suspensions, the group $\mathfrak{F}^\mathbb{Z}$ gives certain connections between the homotopy theory and low dimensional topology.

1.4. The Cohen Construction

Let G be a group and let $\phi \colon G \to [X, \Omega Y]$ be a *representation*, that is, ϕ is a group homomorphism. Write $\mathfrak{K}(X, \Omega Y)$ for the bi-Δ-group given by $\{[X^{n+1}, \Omega Y]\}_{n \geq 0}$ with faces and cofaces described in Section 1.2. By Proposition 1.3.7, the representation ϕ induces a unique morphism of bi-Δ-groups
$$e_\phi \colon \mathfrak{F}^G \longrightarrow \mathrm{sk}_0 \mathfrak{K}(X, \Omega Y) \longrightarrow \mathfrak{K}(X, \Omega Y)$$
and so a homomorphism of progroups
$$e_\phi \colon \mathfrak{H}\mathfrak{F}^G \longrightarrow \mathfrak{H}\mathfrak{K}(X, \Omega Y) = [J(X), \Omega Y].$$

Note that the image of $(\mathfrak{F}^G)_n$ in $[X^{n+1}, \Omega Y]$ is the subgroup generated by
$$d^n d^{n-1} \cdots d^{j+1} d^{j-1} \cdots d^1 d^0(\phi(G)) = p_j^*(\phi(G))$$
for $0 \leq j \leq n$, where $p_j \colon X^{n+1} \to X$ is the $(j+1)$-st coordinate projection.

PROPOSITION 1.4.1. *Let $\phi \colon G \to [X, \Omega Y]$ be a representation. Then the image of*
$$e_\phi \colon \mathfrak{H}\mathfrak{F}^G \longrightarrow \mathfrak{H}\mathfrak{K}(X, \Omega Y) = [J(X), \Omega Y]$$
is given by (possibly infinite) product of the homotopy classes of the following the maps
$$J(X) \xrightarrow{H_n} J(X^{(n)}) \xrightarrow{J(\theta)} J(X^{(m)}) \xrightarrow{J(\phi(g_1) \wedge \cdots \wedge \phi(g_m))} J\left((\Omega Y)^{(m)}\right) \xrightarrow{\tilde{\beta}_m} \Omega Y$$
for $m \geq n \geq 1$, where H_n is the James-Hopf map, g_i runs over elements in G, $\theta \colon X^{(n)} \to X^{(m)}$ runs over the maps $x_1 \wedge \cdots \wedge x_n \mapsto x_{i_1} \wedge x_{i_2} \wedge \cdots \wedge x_{i_m}$ such that $\{i_1, \ldots, i_m\} = \{1, \ldots, n\}$ as sets, $\beta_m \colon (\Omega Y)^{(m)} \to \Omega Y$ runs over all possible m-fold Samelson products and $\tilde{\beta}_m$ is the (unique up to homotopy) H-map extension of β_m.

Note. The map θ occurs as a reduced diagonal if $m > n$ and as a coordinate permutation if $m = n$.

PROOF. By the construction, $(\mathfrak{F}^G)_n = \coprod_{j=0}^{n} (G)_{x_i}$, where $(G)_{x_i}$ is a copy of G. Recall the face operation $d_i \colon (\mathfrak{F}^G)_n \to (\mathfrak{F}^G)_{n-1}$ is a projection map. According to [**46**, Theorem 3.5], $\mathcal{Z}_n \mathfrak{F}^G = \bigcap_{j=0}^{n} \mathrm{Ker}(d_j)$ is the subgroup generated the iterated commutators
$$\beta^t((g_1)_{x_{i_1}}, (g_2)_{x_{i_2}}, \ldots, (g_t)_{x_{i_t}})$$
for $(g_j)_{x_{i_j}} = g_j \in (G)_{x_j}$, where all integers in $\{0, 1, 2, \cdots, n\}$ appear as at least one of the integers i_s and β^t runs over all of the bracket arrangements of iterated commutators of weight $t \geq n+1$. The assertion follows by writing these elements as maps and using the Taylor series (Theorem 1.2.4). □

Consider the representation $e_\phi \colon \mathfrak{F}^\mathbb{Z} \to \mathfrak{K}(X, \Omega\Sigma X)$, where $\phi \colon \mathbb{Z} \to [X, \Omega\Sigma X]$ is given by sending the generator of $\mathbb{Z}$ to the inclusion map $X \subseteq \Omega\Sigma X$. One can check that e_ϕ is a monomorphism if X is an infinite wedge of $\mathbb{C}P^\infty$. However if X or Y runs over certain subclass of spaces, the representations $e_\phi \colon \mathfrak{F}^G \to \mathfrak{K}(X, \Omega Y)$ admit proper kernels and one could get proper quotient bi-Δ-groups of $\mathfrak{F}^G$ as the universal bi-Δ-group model for certain class of representations $\phi \colon G \to [X, \Omega Y]$.

Recall that a space X is called to have *weak LS-category less than k* if the reduced diagonal $\bar{\Delta} \colon X \to X^{(k)}$ is null homotopic. Let $I = (i_1, i_2, \ldots, i_t)$ be a sequence of nonnegative integers with $0 \leq i_s \leq n$ for $1 \leq s \leq t$. Write $l(I)$ for t, the *length* of I. Let
$$r(I) = \max\{\, q \mid i \text{ occurs at least } q \text{ times in } I = (i_1, \ldots, i_t) \text{ for some } 0 \leq i \leq n\}.$$
Consider the map
$$\bar{\pi}_I \colon X^{n+1} \to X^{(t)}, \qquad (z_0, z_1, \ldots, z_n) \mapsto z_{i_1} \wedge z_{i_2} \wedge \cdots \wedge z_{i_t}.$$
The following lemma is useful, which proof is immediate.

LEMMA 1.4.2. *Suppose that X has weak LS-category less than k. If $r(I) \geq k$, then the map $\bar{\pi}_I \colon X^{n+1} \to X^{(t)}$ is null homotopic.* □

A sequence $I = (i_1, \ldots, i_t)$ is called *nondegenerate* if $i_p \neq i_q$ for $1 \leq p < q \leq t$. For convention, the *empty sequence* is regarded as a nondegenerate sequence of length 0. Write **n** for the sequence $(0, 1, \ldots, n)$. Given a nondegenerate sequence $I = (i_1, \ldots, i_t)$, let $\mathbf{n} \smallsetminus I$ denote the nondegenerate monotone increasing sequence obtained from the set $\{0, 1, \ldots, n\} \smallsetminus \{i_1, \ldots, i_t\}$. Recall that the iterated coface $d^{i_1} \circ \cdots \circ d^{i_t} x$ is denoted by $d^I x$, with the convention that $d^I x = x$ if I is the empty sequence.

PROPOSITION 1.4.3. *Let $\mathcal{G} = \mathfrak{K}(X, \Omega Y)$. Suppose that X has weak LS-category less than or equal to k. Then the following statement holds:*

Let $z_i \in \mathcal{Z}_{m_i} \mathcal{G}$ be cycles for $1 \leq i \leq t$ with $0 \leq m_1, \ldots, m_t \leq n$, and let I_i be any nondegenerate monotone decreasing sequence of length $n - m_i$ for $1 \leq i \leq t$. Then the iterated commutator

(1.4.1) $$[[d^{I_1} z_1, d^{I_2} z_2], \ldots, d^{I_t} z_t] = 1$$

for $r(\mathbf{n} \smallsetminus I_1, \mathbf{n} \smallsetminus I_2, \ldots, \mathbf{n} \smallsetminus I_t) \geq k + 1$.

PROOF. Observe that each z_i is represented by the composite

$$X^{m_i+1} \xrightarrow{\text{proj.}} X^{(m_i+1)} \xrightarrow{f_i} \Omega Y$$

for certain map $f_i \colon X^{(m_i+1)} \to \Omega Y$, and the coface $d^{I_i} z_i$ is represented by

$$X^{n+1} \xrightarrow{\bar{\pi}_{\mathbf{n} \smallsetminus I_i}} X^{(m_i+1)} \xrightarrow{f_i} \Omega Y.$$

Thus the element $[[d^{I_1} z_1, d^{I_2} z_2], \ldots, d^{I_t} z_t]$ is represented by the composite

$$X^{n+1} \xrightarrow{\bar{\pi}_J} X^{(m_1+1)} \wedge \cdots \wedge X^{(m_t+1)} \xrightarrow{[[f_1, f_2], \ldots, f_t]} \Omega Y,$$

where $J = (\mathbf{n} \smallsetminus I_1, \ldots, \mathbf{n} \smallsetminus I_t)$. Since $r(J) \geq k + 1$, the assertion follows by Lemma 1.4.2. □

A representation $\phi \colon G \to [X, \Omega Y]$ is called to have *weak LS-category* less than k if the composite

$$X \xrightarrow{\bar{\Delta}_k} X^{(k)} \xrightarrow{[[\phi(g_1), \phi(g_2)], \ldots, \phi(g_k)]} \Omega Y$$

is null homotopic for any $g_1, g_2, \ldots, g_k \in G$. For representations $\phi \colon G \to [X, \Omega Y]$ of weak LS-category $\leq k$, a bi-Δ quotient group of $\mathfrak{F}^G$ is constructed as follows according to the identities in Proposition 1.4.3.

Let $\coprod_{j=0}^{n} (G)_{x_j}$ be the $(n+1)$-fold self free product of G, where $(G)_{x_j}$ is a copy of G. Write $(g)_{x_j}$ for the element $g \in G$ in the j-th copy $(G)_{x_j}$. Define the group $\mathfrak{K}_n^{G,k}$ be the quotient of $\coprod_{j=0}^{n} (G)_{x_j}$ by the relations

(1.4.2) $$[[(g_1)_{x_{i_1}}, (g_2)_{x_{i_2}}], \ldots, (g_t)_{x_{i_t}}] = 1$$

if $r(i_1, i_2, \ldots, i_t) \geq k + 1$. It is straightforward to check that the face and coface operations in $\mathfrak{F}^G$ preserve the above relations. It follows that the sequence of groups $\mathfrak{K}^{G,k} = \{\mathfrak{K}_n^{G,k}\}_{n \geq 0}$ is the quotient bi-Δ-group of $\mathfrak{F}^G$ by the above defining relations.

Note that $\mathfrak{K}^{G,0}$ is the abelianization of $\mathfrak{F}^G$. The first interesting case is $\mathfrak{K}^{G,1}$, which will be simply denoted by $\mathfrak{K}^G$. The groups $\mathfrak{K}_n^G$ (in our notation) has been introduced by Cohen in [9, 10, 11] when $G = \mathbb{Z}$ or $\mathbb{Z}/p^r$. Of course G can be any group as long as there is a group homomorphism $\phi \colon G \to [X, \Omega Y]$ having weak

LS-category less than or equal to 1. This allows to cover many interesting cases. For instance one could consider local spaces with respect to a homology theory. The bi-Δ-group $\mathfrak{K}^{\mathbb{Z}_{(p)},k}$ can be served for p-local spaces. If X or Y is a space with a pointed group Γ-action, G could be chosen to be the group ring $\mathbb{Z}(\Gamma)$.

COROLLARY 1.4.4. *Let $\phi\colon G \to [X, \Omega Y]$ be a representation of weak LS-category less than or equal to k. Then the representation*
$$e_\phi\colon \mathfrak{F}^G \to \mathfrak{K}(X, \Omega Y)$$
factors through the quotient bi-Δ-group $\mathfrak{K}^{G,k}$.

PROOF. From the construction of $\mathfrak{F}^G$, the cofaces are given by
$$d^i(g)_{x_j} = \begin{cases} (g)_{x_j} & \text{for } j < i \\ (g)_{x_{j+1}} & \text{for } j \geq i \end{cases}$$
The assertion follows from the formula
$$[[(g_1)_{x_{i_1}}, (g_2)_{x_{i_2}}], \ldots, (g_t)_{x_{i_t}}]$$
$$= [[d^{(n,\ldots,\hat{i_1},\ldots,0)}(g_1), d^{(n,\ldots,\hat{i_2},\ldots,0)}(g_2)], \ldots, d^{(n,\ldots,\hat{i_t},\ldots,0)}(g_t)],$$
where the sequence $(n, \ldots, \hat{j}, \ldots, 0) = (n, \ldots, j+1, j-1, \ldots, 0)$. $\square$

Note. Observe that the image of $\mathfrak{F}^G$ in $\mathfrak{K}(X, \Omega Y)$ is contained in the bi-Δ-subgroup generated by $\mathcal{Z}_0 \mathfrak{K}(X, \Omega Y)$. Relation (1.4.2) is obtained from Relation (1.4.1) by considering the special cases that all $m_i = 0$. The rest equations in Relation (1.4.1) will be used for considering certain extension of the image of $\mathfrak{F}^G$ in $\mathfrak{K}(X, \Omega Y)$ such as adding possible roots. For instance if the Whitehead square ω_{n+1} is divisible by 2, then the image of $\mathfrak{F}^{\mathbb{Z}}$ in $\mathfrak{K}(S^n, \Omega S^{n+1})$ admits a canonical extension by adding $\frac{1}{2}\omega_{n+1} \in \mathcal{Z}_1 \mathfrak{K}(S^n, \Omega S^{n+1})$.

Write $\{\Gamma^n G\}$ for the descending series of a group G starting with $\Gamma^1 G = G$.

PROPOSITION 1.4.5. *Let G be any group. Then*
1) $\Gamma^{(n+1)k+1} \mathfrak{K}_n^{G,k} = \{1\}$ *for $n \geq 0$. Thus $\mathfrak{K}^{G,k}$ is a dimension-wise nilpotent bi-Δ-group.*
2) *The cycles $\mathcal{Z}_n \mathfrak{K}^G = \Gamma^{n+1} \mathfrak{K}_n^G$ for each $n \geq 0$.*

PROOF. Assertion (1) follows immediately from Relation 1.4.2.

(2). From (1), $\Gamma^{n+2} \mathfrak{K}_n^G = 1$ for each $n \geq 0$ and so
$$\Gamma^{n+1} \mathfrak{K}_n^G \subseteq \mathcal{Z}_n \mathfrak{K}^G.$$
By the proof of Proposition 1.4.1,
$$\mathcal{Z}\mathfrak{F}^G = \bigcap_{j=0}^n \mathrm{Ker}(d_j) \subseteq \Gamma^{n+1} \mathfrak{F}_n^G.$$
Thus
$$\mathcal{Z}\mathfrak{K}^G \subseteq \Gamma^{n+1} \mathfrak{K}_n^G$$
by Proposition 1.2.10 and hence the result. $\square$

A relation between the descending central series of $\mathfrak{K}^G$ and the Postnikov system defined in Theorem 1.2.7 is as follows.

PROPOSITION 1.4.6. *Let $\{P_n\mathfrak{K}^G\}$ be the Postnikov tower of $\mathfrak{K}^G$. Then*
$$P_n\mathfrak{K}^G \cong \mathfrak{K}^G/\Gamma^{n+2}\mathfrak{K}^G$$
as bi-Δ-groups.

PROOF. Since $\mathcal{Z}_t\mathfrak{K}^G \subseteq \Gamma^{n+2}\mathfrak{K}^G$ for each $t \geq n+1$, the composite
$$\mathcal{Z}R_n\mathfrak{K}^G \to \mathcal{Z}\mathfrak{K}^G \to \mathcal{Z}(\mathfrak{K}^G/\Gamma^{n+2}\mathfrak{K}^G)$$
is trivial and, by Proposition 1.2.9, so is the composite $R_n\mathfrak{K}^G \to \mathfrak{K}^G \to \mathfrak{K}^G/\Gamma^{n+2}\mathfrak{K}^G$. It follows that the quotient map $\mathfrak{K}^G \to \mathfrak{K}^G/\Gamma^{n+2}\mathfrak{K}^G$ factors through $P_n\mathfrak{K}^G$. Now consider the composite
$$\phi\colon \Gamma^{n+2}\mathfrak{K}^G \longrightarrow \mathfrak{K}^G \longrightarrow P_n\mathfrak{K}^G.$$
By Proposition 1.2.10, $\mathcal{Z}\Gamma^{n+2}\mathfrak{K}^G = \Gamma^{n+2}\mathfrak{K}^G \cap \mathcal{Z}\mathfrak{K}^G$ and so $\mathcal{Z}_t\Gamma^{n+2}\mathfrak{K}^G = 1$ for $t \leq n$. By Theorem 1.2.7, $\mathcal{Z}_t P_n\mathfrak{K}^G = 1$ for $t > n$. Thus $\mathcal{Z}(\phi)\colon \mathcal{Z}\Gamma^{n+2}\mathfrak{K}^G \to \mathcal{Z}P_n\mathfrak{K}^G$ is trivial and so is ϕ. It follows that the quotient $\mathfrak{K}^G \to P_n\mathfrak{K}^G$ factors through $\mathfrak{K}^G/\Gamma^{n+2}\mathfrak{K}^G$ and hence the result. □

Note. This proposition does not hold for $\mathfrak{K}^{G,k}$ with $k > 1$.

PROPOSITION 1.4.7. *Let G be any group. Then the canonical quotient*
$$\mathfrak{K}^{G,k} \to \mathfrak{K}^{G/\Gamma^{k+1}G,k}$$
is an isomorphism of bi-Δ-groups. In particular, $\mathfrak{K}^G \to \mathfrak{K}^{G^{\mathrm{ab}}}$ is an isomorphism.

PROOF. By the defining relations, the composite
$$\overbrace{[[(G)_{x_i},(G)_{x_i}],\ldots,(G)_{x_i}]}^{t} \hookrightarrow \coprod_{j=0}^{n}(G)_{x_j} \twoheadrightarrow \mathfrak{K}_n^{G,k}$$
is trivial if the commutator length $t > k$ and so the quotient map $\mathfrak{F}^G \to \mathfrak{K}^{G,k}$ factors through $\mathfrak{F}^{G/\Gamma^{k+1}G}$. Moreover the resulting quotient $\mathfrak{F}^{G/\Gamma^{k+1}G} \to \mathfrak{K}^{G,k}$ factors through $\mathfrak{K}^{G/\Gamma^{k+1}G,k}$ and hence the result. □

Next we are going to determine the descending central series of the group $\mathfrak{K}_n^G$ and we need to introduce some constructions that are essentially from the ideas given in [**9**] but are set up in more general and systematic situation. Let M be any abelian group regarded as a module over $\mathbb{Z}$ and so the tensor product will be given over integers. Let V_n be the $(n+1)$-fold direct sum of M. Let
$$M_i = \{(0,\ldots,0,x,0,\ldots,0) \in V_n | x \in M\}$$
for $0 \leq i \leq n$. Observe that $V_n = \sum_{i=0}^{n} M_i$ is a direct sum. Let $\mathfrak{A}_n^M$ be the quotient algebra of the tensor algebra $T(V_n)$ modulo the two sided ideal generated by the homogeneous elements in
$$M_{i_1} \otimes M_{i_2} \otimes \cdots \otimes M_{i_t}$$
with $i_a = i_b$ for some $1 \leq a < b \leq t$. As a graded group,
$$(\mathfrak{A}_n^M)_t = \bigoplus_{\substack{0 \leq i_1 < i_2 < \cdots < i_t \leq n \\ \sigma \in S_t}} M_{i_{\sigma(1)}} \otimes M_{i_{\sigma(2)}} \otimes \cdots \otimes M_{i_{\sigma(t)}}.$$

The elements in M_i is denoted by αy_i with $\alpha \in M$ under the canonical isomorphism $M_i \cong M$. Observe that $\alpha y_i + \beta y_i = (\alpha + \beta) y_i$ in V_n and any element α in V_n has a unique decomposition $\alpha = \sum_{i=0}^{n} \alpha_i y_i$ for $\alpha_i \in M$. The relations in $\mathfrak{A}_n^M$ can be rewritten as
$$(\alpha_1 y_{i_1})(\alpha_2 y_{i_2}) \cdots (\alpha_t y_{i_t}) = 0$$
if $i_a = i_b$ for some $1 \leq a < b \leq t$.

Let $d_i \colon V_n \to V_{n-1}$ be the linear map such that
$$d_i(\alpha y_j) = \begin{cases} \alpha y_j & \text{for } j < i \\ 0 & \text{for } j = i \\ \alpha y_{j-1} & \text{for } j > i \end{cases}$$
and let $d^i \colon V_n \to V_{n+1}$ be the linear map such that
$$d^i(\alpha y_j) = \begin{cases} \alpha y_j & \text{for } j < i \\ \alpha y_{j+1} & \text{for } j \geq i. \end{cases}$$

Then the sequence of groups $\{V_n\}_{n \geq 0}$ is an abelian bi-Δ-group with the above faces and cofaces. Moreover $\{\mathfrak{A}_n^M\}_{n \geq 0}$ is a bi-Δ graded algebra with faces and cofaces induced from that on $\{V_n\}_{n \geq 0}$.

Let $G_{x_i} = \{1 + \alpha y_i | \alpha \in M\} \subseteq \mathfrak{A}_n^M$. Since $(1 + \alpha y_i)(1 - \alpha y_i) = 1$ in $\mathfrak{A}_n(M)$ for $\alpha \in M_i$, G_{x_i} is contained in the group of units $(\mathfrak{A}_n^M)^*$ of $\mathfrak{A}_n^M$ and so the inclusions $G_{x_i} \hookrightarrow (\mathfrak{A}_n^M)^*$ extends uniquely to a group homomorphism
$$e \colon \coprod_{i=1}^{n} G_{x_i} \longrightarrow (\mathfrak{A}_n^M)^* \subseteq \mathfrak{A}_n^M.$$

Observe that as an abelian group $G_{x_i} \cong M$. The above group homomorphism induces a representation $e \colon \mathfrak{F}^M \to \mathfrak{A}^M$.

LEMMA 1.4.8. *Let* $e \colon \mathfrak{F}^M \to \mathfrak{A}^M$ *be the representation defined as above. Then*
$$e([[(g_1)_{x_{i_1}}, (g_2)_{x_{i_2}}], \cdots, (g_t)_{x_{i_t}}]) = 1 + [[g_1 y_{i_1}, g_2 y_{i_2}], \cdots, g_t y_{i_t}]$$
for $g_1, g_2, \ldots, g_t \in M$, *where the bracket on the left side is the commutator in the group* $\coprod_{i=1}^{n} G_{x_i}$ *and the bracket on the right side is the commutator in the algebra* $\mathfrak{A}_n(M)$.

PROOF. The proof is given by induction on t. When $t = 2$, we have
$$e([(g_1)_{x_{i_1}}, (g_2)_{x_{i_2}}]) = e((g_1)_{x_{i_1}})^{-1} e((g_2)_{x_{i_2}})^{-1} e((g_1)_{x_{i_1}}) e((g_2)_{x_{i_2}})$$
$$= (1 - g_1 y_{i_1})(1 - g_2 y_{i_2})(1 + g_1 y_{i_1})(1 + g_2 y_{i_2}) = 1 + [g_1 y_{i_1}, g_2 y_{i_2}].$$
Suppose that the assertion holds for $t - 1$ with $t \geq 3$. Let
$$w = [[(g_1)_{x_{i_1}}, (g_2)_{x_{i_2}}], \cdots, (g_{t-1})_{x_{i_{t-1}}}].$$
By induction, $e(w) = 1 + [[g_1 y_{i_1}, g_2 y_{i_2}], \cdots, g_{t-1} y_{i_{t-1}}]$ and so
$$e(w)^{-1} = 1 - [[g_1 y_{i_1}, g_2 y_{i_2}], \cdots, g_{t-1} y_{i_{t-1}}].$$
It follows that
$$e([w, (g_t)_{x_{i_t}}]) = e(w)^{-1}(1 - g_t y_{i_t}) e(w)(1 + g_t y_{i_t}) = 1 + [[g_1 y_{i_1}, g_2 y_{i_2}], \cdots, g_t y_{i_t}]$$
and hence the result. $\square$

1.4. THE COHEN CONSTRUCTION

COROLLARY 1.4.9. *Let M be any abelian group. Then representation $e\colon \mathfrak{F}^M \to \mathfrak{A}^M$ induces a representation $e\colon \mathfrak{K}^M \to \mathfrak{A}^M$.* □

Define $\widetilde{\mathrm{Lie}}^M(n+1)$ to the sub-$\mathbb{Z}$-module of $(V_n)^{\otimes n}$ spanned by the Lie elements

$$[[g_0 y_{\sigma(0)}, g_1 y_{\sigma(1)}], \ldots, (g_n) y_{\sigma(n)}]$$

for $g_i \in M$ and $\sigma \in S_{n+1}$ acting on $\{0, 1, \ldots, n\}$.

THEOREM 1.4.10. *Let G be any group. Then*

(1). *There is an isomorphism of groups*

$$\Gamma^{n+1} \mathfrak{K}_n^G \cong \widetilde{\mathrm{Lie}}^{G^{\mathrm{ab}}}(n+1)$$

for each $n \geq 0$.

(2). *There is a decomposition:*

$$\Gamma^{k+1} \mathfrak{K}_n^G / \Gamma^{k+2} \mathfrak{K}_n^G \cong \bigoplus_{0 \leq i_1 < i_2 < \cdots < i_{n-k} \leq n} d^{i_{n-k}} d^{i_{n-k-1}} \cdots d^{i_1} (\Gamma^{k+1} \mathfrak{K}_k^G)$$

for $0 \leq k \leq n-1$.

PROOF. (1). By Proposition 1.4.7, we may assume that G is abelian. Consider the representation $e\colon \mathfrak{K}_n^G \to \mathfrak{A}_n^G$. By Lemma 1.4.8, the map e induces a group homomorphism

$$\bar{e}\colon \Gamma^{n+1} \mathfrak{K}_n^G \to 1 + \widetilde{\mathrm{Lie}}^G(n+1) \cong \widetilde{\mathrm{Lie}}^G(n+1).$$

Now the quotient map $\coprod_{j=0}^n (G)_{x_j} \to \mathfrak{K}_n^G$ induces an epimorphism

$$\theta\colon \Gamma^{n+1} \left(\coprod_{j=0}^n (G)_{x_j} \right) / \Gamma^{n+2} \left(\coprod_{j=0}^n (G)_{x_j} \right) \twoheadrightarrow \Gamma^{n+1} \mathfrak{K}_n^G.$$

By taking iterated Lie operation, there is a canonical map

$$j\colon \widetilde{\mathrm{Lie}}^G(n+1) \longrightarrow \Gamma^{n+1} \left(\coprod_{j=0}^n (G)_{x_j} \right) / \Gamma^{n+2} \left(\coprod_{j=0}^n (G)_{x_j} \right).$$

By the defining relations for $\mathfrak{K}_n^G$, the composite $\theta \circ j\colon \widetilde{\mathrm{Lie}}^G(n+1) \to \Gamma^{n+1} \mathfrak{K}_n^G$ is an epimorphism. Clearly the composite $\bar{e} \circ \theta \circ j\colon \widetilde{\mathrm{Lie}}^G(n+1) \to \mathrm{Lie}^G(n+1)$ is the identity map. Thus $\bar{e}\colon \Gamma^{n+1} \mathfrak{K}_n^G \to \widetilde{\mathrm{Lie}}^G(n+1)$ is an isomorphism and assertion (1) follows.

(2). Consider the short exact sequence of bi-Δ-groups

$$\Gamma^{n+1} \mathfrak{K}^G / \Gamma^{n+2} \mathfrak{K}^G \hookrightarrow \mathfrak{K}^G / \Gamma^{n+2} \mathfrak{K}^G \twoheadrightarrow \mathfrak{K}^G / \Gamma^{n+1} \mathfrak{K}^G.$$

By Theorem 1.2.7 and Proposition 1.4.6, the bi-Δ-group $\Gamma^{n+1}\mathfrak{K}^G/\Gamma^{n+2}\mathfrak{K}^G$ has only cycles in dimension n. Since $\Gamma^{n+1}\mathfrak{K}^G/\Gamma^{n+2}\mathfrak{K}^G$ is an abelian bi-Δ-group, the semi-direct product decomposition in Proposition 1.2.9 is a product decomposition that gives the decomposition formula in assertion (2). This finishes the proof. □

Note. If R is a subring of $\mathbb{Q}$ or $R = \mathbb{Z}/p^r$, then $R^{\otimes n} = R$ and, in this case, $\widetilde{\mathrm{Lie}}^R(n+1)$ is the usual $\mathrm{Lie}^R(n+1)$, that is, the sub-R-module of $(Ry_0 \oplus \cdots Ry_n)^{\otimes n}$ spanned by the Lie elements

$$[[y_{\sigma(0)}, y_{\sigma(1)}], \ldots, y_{\sigma(n)}]$$

for $\sigma \in S_{n+1}$, where y_i is identified with $1y_i$. For a general abelian group M, clearly

$$\widetilde{\mathrm{Lie}}^M(n+1) \cong \mathrm{Lie}^{\mathbb{Z}}(n+1) \otimes M^{\otimes n+1}$$

as modules over $\mathbb{Z}(S_{n+1})$, where S_{n+1} acts on $M^{\otimes n+1}$ by permuting positions. In particular, for any group Γ, $\widetilde{\mathrm{Lie}}^{\mathbb{Z}_{(p)}(\Gamma)}(n+1)$ is a module over the wreath product $S_{n+1} \wr \Gamma$.

COROLLARY 1.4.11. *Let M be any abelian group. Then $e \colon \mathfrak{K}^M \to \mathfrak{A}^M$ is a faithful representation.*

PROOF. By Proposition 1.4.5 and Theorem 1.4.10,

$$e \colon \mathcal{Z}_n \mathfrak{K}^M = \Gamma^{n+1} \mathfrak{K}_n^M \to \mathcal{Z}_n \mathfrak{A}^M \subseteq \mathfrak{A}_n^M$$

is faithful for all n. The assertion follows from Proposition 1.2.10. □

CHAPTER 2

Shuffle Relations

2.1. Functors to Coalgebras

In this section, the ground ring R is a commutative ring with identity. A coalgebra E means a pointed (ungraded or graded) coassociative cocommutative R-coalgebra. Write $\eta\colon R \to E$ and $\epsilon\colon E \to R$ for the coaugmentation and counit, respectively. A *functor to coalgebras* means a functor C from a (fixed) category to the category of pointed coassociative cocommutative coalgebras over R. Write $\eta\colon R \to C$ and $\epsilon\colon C \to R$ for the coaugmentation and counit transformation, respectively, where R is the constant functor to coalgebras by sending each object to the trivial coalgebra R. A natural transformation from C to D is called a *coalgebra map*. The class of natural transformations from C to D is denoted by $\mathrm{coalg}(C, D)$. A natural transformation from C to D as functors to R-modules is called an *R-homomorphism*. Write $\mathrm{Hom}_R(C, D)$ for the class of natural transformations from C to D as functors to R-modules. A pair of functors (C, D) to coalgebras is called *relatively small* if the class of natural transformation from C to D is a set. For instance, the class $\mathrm{coalg}(C, D)$ is a set if 1) (C, D) is a pair of functors from a small category $\mathcal{C}$ to coalgebras or 2) C preserves direct limits and every object in $\mathcal{C}$ is a direct limit of objects over a small subcategory of $\mathcal{C}$. In this article we always assume that a pair of functors (C, D) to coalgebras is relatively small and so $\mathrm{coalg}(C, D)$ and $\mathrm{Hom}_R(C, D)$ will be always considered as sets.

The coproduct in the category of functors to coalgebras is denoted by $C \vee D$. Note that $C \vee D$ is the push-out of the diagram $C \xleftarrow{\eta} R \xrightarrow{\eta} D$. We also call $C \vee D$ is the *wedge* of C and D. The product of pointed coalgebras is the usual tensor product $C \otimes D$. The *smash product* is defined to be the cokernel of the inclusion $C \vee D \to C \otimes D$ in the category of functors to coalgebras. The *reduced diagonal* $\bar{\psi}\colon C \to C \wedge C$ is defined to be the composite $C \xrightarrow{\psi} C \otimes C \twoheadrightarrow C \wedge C$. The *primitive functor* of C is defined to be $PC = \mathrm{Ker}(\bar{\psi}\colon C \to C \wedge C)$ as functors to modules.

A coalgebra A is called a *quasi-Hopf algebra* if A has a coalgebra multiplication $\mu\colon A \otimes A \to A$. In other words, a quasi-Hopf algebra is an H-object in the category of pointed coalgebras. We always assume that a quasi-Hopf algebra is coassociative and cocommutative. According to [**30**], a *Hopf algebra* means an associative quasi-Hopf algebra. Let A be a quasi-Hopf algebra. The R-module $Q(A) = IA/(IA \cdot IA)$ is called the set of indecomposable elements.

Let C be a functor to coalgebras and let A be a functor to quasi-Hopf algebras. The *convolution product* $f * g$ is defined by

$$C \xrightarrow{\psi} C \otimes C \xrightarrow{f \otimes g} A \otimes A \xrightarrow{\mu} A$$

for $f, g\colon C \to A$. Note that $\mathrm{coalg}(C, A)$ is an H-set under the convolution product (see [**30**]). If A is a functor to Hopf algebras (with conjugation), then $\mathrm{coalg}(C, A)$ is monoid (group).

Let F be a functor to coalgebras. Recall that the *cobar construction* $\mathrm{Cobar}(F)$ is a cosimplicial set defined by
$$\mathrm{Cobar}(F)_n = F^{\otimes n}$$
with cofaces d^i and codegeneracies s^i for $0 \leq i \leq n$ given by
$$d^0 = \eta \otimes \mathrm{id}\colon \mathrm{Cobar}(F)_{n-1} = F^{\otimes n-1} = R \otimes F^{\otimes n-1} \hookrightarrow F^{\otimes n} = \mathrm{Cobar}(F)_n,$$
$$d^n = \mathrm{id} \otimes \eta\colon \mathrm{Cobar}(F)_{n-1} = F^{\otimes n-1} = F^{\otimes n-1} \otimes R \hookrightarrow F^{\otimes n} = \mathrm{Cobar}(F)_n,$$
$$d^i\colon \mathrm{Cobar}(F)_{n-1} = F^{\otimes n-1} \xrightarrow{\mathrm{id}_{F^{\otimes i-1}} \otimes \psi \otimes \mathrm{id}_{F^{\otimes n-i-1}}} F^{\otimes n} = \mathrm{Cobar}(F)_n,$$
$$s^i\colon \mathrm{Cobar}(F)_{n+1} = F^{\otimes n+1} \xrightarrow{\mathrm{id}_{F^{\otimes i}} \otimes \epsilon \otimes \mathrm{id}_{F^{\otimes n-i}}} F^{\otimes i} \otimes R \otimes F^{n-i} = \mathrm{Cobar}(F)_n.$$
Clearly $\mathrm{Cobar}(F)$ is cosimplicial functor to coalgebras. Let D be a functor to coalgebras. Define the simplicial set $\mathcal{W}(F, D)$ by
$$\mathcal{W}(F, D)_n = \mathrm{coalg}(\mathrm{Cobar}(F)_n, D) = \mathrm{coalg}(F^{\otimes n}, D)$$
with faces and degeneracies induced by the cofaces and codegeneracies on $\mathrm{Cobar}(F)$. Note that $\mathcal{W}(F, D)_1 = \mathrm{coalg}(F, D)$. The canonical function
$$\sigma\colon \mathrm{coalg}(F, D) \to H_1(\mathcal{W}(F, D); \mathbb{Z})$$
is called the *suspension*.

Remark. Recall that [**14, 24**] any simplicial set X is a contravariant functor from ordered finite sets to sets, and $R(X)$ is a functor from the opposite category of ordered finite sets to coalgebras. The cobar construction for simplicial sets is used for constructing loop spaces. The function from $\mathrm{coalg}(F, D)$ to the fundamental group of $\mathcal{W}(F, D))$ seems the algebraic analogue of the function $\Omega\colon [X, Y] \to [\Omega X, \Omega Y]$.

LEMMA 2.1.1. *Let F and A be functors to coalgebras.*
1) *If A is a functor to Hopf algebras, then $\mathcal{W}(F, A)$ is a simplicial monoid;*
2) *If A is a functor to Hopf algebras with conjugation, then $\mathcal{W}(F, A)$ is a simplicial group;*
3) *If A is a functor to quasi-Hopf algebras with conjugation, then $\mathcal{W}(F, A)$ is a fibrant simplicial H-set.*

PROOF. Assertions (1) and (2) are obvious. Assertion (3) follows from the fact that (i) any functor to quasi-Hopf algebras (with conjugation) is a coalgebra retract of a functor to Hopf algebras (with conjugation) and (ii) any retract of a fibrant simplicial set is fibrant. □

Let $\mathfrak{R}(F, A)$ denote $H_1(\mathcal{W}(F, A); \mathbb{Z})$.

PROPOSITION 2.1.2. *Let F and F' be functors to coalgebras and let A and A' be functors to quasi-Hopf algebras with conjugation. Then the following statements hold:*
1) *There is a right short exact sequence of H-sets*
$$\mathrm{coalg}(F \wedge F, A) \xrightarrow{\bar{\psi}^*} \mathrm{coalg}(F, A) \longrightarrow \mathfrak{R}(F, A) \longrightarrow 1.$$
2) *$\mathfrak{R}(A, A))$ admits a ring structure under the composition of self coalgebra natural transformations of A.*

3) *There is a decomposition*

$$\mathfrak{R}(F \otimes F', A \otimes A') \cong \mathfrak{R}(F, A) \oplus \mathfrak{R}(F, A') \oplus \mathfrak{R}(F', A) \oplus \mathfrak{R}(F', A').$$

PROOF. (1). Since $\mathcal{W}(F, A)_0 = \{1\}$, $\operatorname{coalg}(F, A) = \mathcal{W}(F, A)_1$ is the set of 1-dimensional cycles. By Lemma 2.1.1, $\mathcal{W}(F, A)$ is a fibrant H-set. It follows that

$$\pi_1(\mathcal{W}(F, A)) \cong H_1(\mathcal{W}(F, A); \mathbb{Z})$$

and

$$\operatorname{coalg}(F, A) = \mathcal{W}(F, A)_1 \longrightarrow \pi_1(\mathcal{W}(F, A)) \cong \mathfrak{R}(F, A)$$

is an epimorphism of H-sets with the kernel given by

$$d_1 \left(\operatorname{Ker}(d_0 \colon \mathcal{W}(F, A)_2 \to \mathcal{W}(F, A)_1) \cap \operatorname{Ker}(d_2 \colon \mathcal{W}(F, A)_2 \to \mathcal{W}(F, A)_1) \right)$$

in $\mathcal{W}(F, A)_1$. Assertion (1) follows.

(2). Any coalgebra map $h \colon A \to A$ induces simplicial maps

$$h_*, h^* \colon \mathcal{W}(A, A) \to \mathcal{W}(A, A),$$

which give the group homomorphisms

$$h_*, h^* \colon \mathfrak{R}(A, A) = H_1(\mathcal{W}(A, A); \mathbb{Z}) \to \mathfrak{R}(A, A) = H_1(\mathcal{W}(A, A); \mathbb{Z})$$

and hence Assertion (2).

(3). Since $\mathcal{W}(F \otimes F', A \otimes A') \cong \mathcal{W}(F \otimes F', A) \times \mathcal{W}(F \otimes F', A')$, there is a decomposition

$$\mathfrak{R}(F \otimes F', A \otimes A') \cong \mathfrak{R}(F \otimes F', A) \oplus \mathfrak{R}(F \otimes F', A').$$

It suffices to show that

$$\mathfrak{R}(F \otimes F', A) \cong \mathfrak{R}(F, A) \oplus \mathfrak{R}(F', A).$$

There is a functor B to Hopf algebras with conjugation such that A is a coalgebra retract of B. The inclusions

$$i_1 \colon F = F \otimes R \to F \otimes F' \text{ and } i_2 \colon F' = R \otimes F' \to F \otimes F'$$

induce a homomorphism of simplicial groups

$$(i_1, i_2)^* \colon \operatorname{coalg}(\operatorname{Cobar}(F \otimes F'), B) \longrightarrow \operatorname{coalg}(\operatorname{Cobar}(F), B) \times \operatorname{coalg}(\operatorname{Cobar}(F'), B).$$

Since B is a functor to Hopf algebras, any coalgebra map $F^{\otimes n} \vee F'^{\otimes n} \to B$ extends to $(F \otimes F')^{\otimes n}$ and so

$$(i_1, i_2)^* \colon \mathcal{W}(F \otimes F', B) \longrightarrow \mathcal{W}(F, B) \times \mathcal{W}(F', B)$$

is an epimorphism of simplicial groups. Let $\mathcal{G}$ denote the kernel of $(i_1, i_2)^*$. There is an exact sequence of groups

$$\pi_1(\mathcal{G}) \xrightarrow{j_*} \pi_1(\mathcal{W}(F \otimes F', B)) \longrightarrow \pi_1(\mathcal{W}(F, B) \times \mathcal{W}(F', B)) \longrightarrow \pi_0(\mathcal{G}) = 1.$$

Note that the group $\mathcal{G}_1$ consists of coalgebra map $\phi \colon F \otimes F' \to B$ such that ϕ restricted to $F \vee F'$ is trivial, that is, ϕ factors through $F \wedge F'$. Since the quotient map $q \colon F \otimes F' \to F \wedge F'$ is given by the composite

$$F \otimes F' \xrightarrow{\bar\psi} (F \otimes F') \wedge (F \otimes F') \xrightarrow{(F \otimes \epsilon) \wedge (\epsilon \otimes F')} F \wedge F',$$

the group G_1 is contained in

$$\operatorname{Im}(\bar\psi \colon \operatorname{coalg}((F \otimes F') \wedge (F \otimes F'), B) \to \operatorname{coalg}(F \otimes F', B)).$$

Thus
$$j_*\colon \pi_1(\mathcal{G}) \to \pi_1(\mathcal{W}(F \otimes F', B))$$
is trivial and so the map $(i_1, i_2)^*$ from $\pi_1(\mathcal{W}(F \otimes F', B))$ to
$$\pi_1(\mathcal{W}(F, B) \times \mathcal{W}(F', B)) \cong \pi_1(\mathcal{W}(F, B)) \times \pi_1(\mathcal{W}(F', B))$$
is an isomorphism. This gives the decomposition
$$(i_1, i_2)^*\colon \mathfrak{R}(F \otimes F', B) \cong \mathfrak{R}(F, B) \oplus \mathfrak{R}(F', B).$$
From the commutative diagram

$$\begin{array}{ccc}
\mathfrak{R}(F \otimes F', A) & \xrightarrow{(i_1, i_2)^*} & \mathfrak{R}(F, A) \oplus \mathfrak{R}(F', A) \\
\uparrow & & \uparrow \\
\mathfrak{R}(F \otimes F', B) & \xrightarrow[\cong]{(i_1, i_2)^*} & \mathfrak{R}(F, B) \oplus \mathfrak{R}(F', B) \\
\downarrow & & \downarrow \\
\mathfrak{R}(F \otimes F', A) & \xrightarrow{(i_1, i_2)^*} & \mathfrak{R}(F, A) \oplus \mathfrak{R}(F', A),
\end{array}$$

the map $(i_1, i_2)^*\colon \mathfrak{R}(F \otimes F', A) \to \mathfrak{R}(F, A) \oplus \mathfrak{R}(F', A)$ is an isomorphism and hence the result. $\square$

Remark. Some remarks to Proposition 2.1.2 are given:

1) Let A be a functor to quasi-Hopf algebras and let C and C' be functors to coalgebras. Then
$$(f * g) \circ h = (f \circ h) * (g \circ h)\colon C' \to A$$
for $f, g \in \mathrm{coalg}(C, A)$ and $h \in \mathrm{coalg}(C', C)$. In particular, one-side distributivity law already hold in $\mathrm{coalg}(A, A)$ but another side distributivity law fails in general. Recall that an H-set X is called a *semiring* if there is a binary operation in X such that one-side distributivity law with respect to the multiplication. Thus $\mathrm{coalg}(A, A)$ is a semiring and the suspension $\sigma\colon \mathrm{coalg}(A, A) \to \mathfrak{R}(A, A)$ is a morphism of semirings.

2) Let A be a functor to quasi-Hopf algebras with conjugation that admits a coalgebra decomposition $A \cong A_1 \otimes A_2$, that is, there are functors to coalgebras A_1 and A_2 such that $A \cong A_1 \otimes A_2$ as functors to coalgebras. By assertion (3), there is a decomposition

$$\mathfrak{R}(A, A) \cong \mathfrak{R}(A_1 \otimes A_2, A_1 \otimes A_2) \cong \mathfrak{R}(A_1, A_1) \oplus \mathfrak{R}(A_1, A_2) \oplus \mathfrak{R}(A_2, A_1) \oplus \mathfrak{R}(A_2, A_2).$$

Thus the coalgebra decompositions of the functor A are one-to-one correspondent to the decompositions of the identity in the ring $\mathfrak{R}(A, A)$ in terms of idempotents. In other words, the study on coalgebra decompositions of the functor A is equivalent to the representation theory of the ring $\mathfrak{R}(A, A)$. $\square$

Let F be a functor to pointed coalgebras. A *bi-Δ-resolution* of F is a sequence of functors to coalgebras $\{F^{\otimes n+1}\}_{n \geq 0}$ with faces and cofaces given by
$$d_i = \mathrm{id}_F^{\otimes i} \otimes \epsilon \otimes \mathrm{id}_F^{n-i}\colon F^{\otimes n+1} \to F^{\otimes n},$$

2.1. FUNCTORS TO COALGEBRAS

$$d^i = \mathrm{id}_F^{\otimes i} \otimes \eta \otimes \mathrm{id}_F^{n-i} \colon F^{\otimes n} \to F^{\otimes n+1}$$

for $0 \le i \le n$. Let D be a functor to pointed coalgebras. Then the sequence of sets

$$\mathfrak{K}(F, D) = \{\mathrm{coalg}(F^{\otimes n+1}, D)\}_{n \ge 0}$$

is a bi-Δ-set with faces $d_i = d^{i*}$ and $d^i = d_i^*$. If D is a functor to quasi-Hopf algebras (Hopf algebras with conjugation), then $\mathfrak{K}(F, D)$ is a bi-Δ H-set (bi-Δ-group). We are going to consider some special but important functors.

Let V be a (ungraded) projective R-module. The tensor algebra $T(V)$ is a Hopf algebra by saying V primitive. This gives a functor T from projective R-modules to Hopf algebras with conjugation. Write $T_n(V)$ for $V^{\otimes n}$. Let

$$J_n(V) = \bigoplus_{j=0}^{n} T_j(V).$$

Then $J_n(V)$ is a subcoalgebra of $T(V)$ and so J_n is a functor from projective R-modules to pointed coalgebras. Note that $J_1(V) = R \oplus V$ with trivial comultiplication.

Let $P^n(p^r)$ be the n-dimensional mod p^r Moore space and let $\Omega(P^3(p^r) \wedge -)$ denote the functor $X \mapsto \Omega(P^3(p^r) \wedge X)$. Let $j = j \wedge \mathrm{id}_X \colon S^2 \wedge X \to P^3(p^r) \wedge X$ be the inclusion map. The proof of the following theorem follows the lines in [**39**, Sections 2-3].

THEOREM 2.1.3. *The following statements hold.*

1) *The natural transformations between T_n and T_m as functors to R-modules are given by*

$$\mathrm{Hom}_R(T_n, T_m) = \begin{cases} 0 & \text{if } n \ne m \\ R(S_n) & \text{if } n = m, \end{cases}$$

   where $R(S_n)$ is the group ring of the symmetric group with coefficients in the ring R.

2) *The inclusion $i_n \colon J_{n-1} \to J_n$ induces an epimorphism of groups*

$$i_n^* \colon \mathrm{coalg}(J_n, T) \twoheadrightarrow \mathrm{coalg}(J_{n-1}, T)$$

   with

$$\mathrm{Ker}(i_n^* \colon \mathrm{coalg}(J_n, T) \to \mathrm{coalg}(J_{n-1}, T)) \cong \mathrm{Lie}^R(n).$$

   for each n.

3) *The group $\mathrm{coalg}(T, T)$ is given by the inverse limit*

$$\mathrm{coalg}(T, T) \cong \varprojlim_{i_n^*} \mathrm{coalg}(J_n, T).$$

4) *There is an isomorphism of progroups*

$$\mathrm{coalg}(T, T) \cong \mathfrak{H}\mathfrak{K}(J_1, T).$$

5) *If $R = \mathbb{Z}, \mathbb{Z}_{(p)}$ or $\mathbb{Q}$, then the composite*

$$\mathfrak{H}\mathfrak{K}^R \xrightarrow{e_\phi} [\Omega\Sigma^2, \Omega\Sigma^2] \xrightarrow{H_*} \mathrm{coalg}^{R,\mathrm{graded}}(T, T) \xrightarrow{\mathrm{restriction}} \mathrm{coalg}(T, T)$$

   is an isomorphism, where the bi-Δ-group $\mathfrak{K}^R$ is the quotient of $\mathfrak{F}^R$ defined before Corollary 1.4.4 and the map e_ϕ is given in Proposition 1.4.1.

6) If $R = \mathbb{Z}/p^r$, then the composite

$$\mathfrak{H}\mathfrak{K}^R \xrightarrow{e_j} [\Omega\Sigma^2, \Omega(P^3(p^r) \wedge -)] \xrightarrow{H_*} \mathrm{coalg}^{R,\mathrm{graded}}(T,T) \xrightarrow{\mathrm{restriction}} \mathrm{coalg}(T,T)$$

is an isomorphism, where the map e_j is given in Proposition 1.4.1. □

REMARK 2.1.4. We give some remarks to Theorem 2.1.3:

(1). By assertion (1), the inclusion $\mathrm{coalg}(J_n, J_n) \to \mathrm{coalg}(J_n, T)$ is an isomorphism and so $\mathrm{coalg}(J_n, J_n)$ admits a group structure induced from $\mathrm{coalg}(J_n, T)$. Moreover under the convolution product $\mathrm{coalg}(J_n, J_n)$ is a semiring.

(2). For any commutative ring R with identity, by using the universal property of $\mathfrak{F}^R$, there is a (unique) morphism of bi-Δ-groups from $\mathfrak{F}^R$ to $\mathfrak{K}(J_1, T)$ that factors through the quotient bi-Δ-group $\mathfrak{K}^R$. The induced morphism of progroups

$$\mathfrak{H}\mathfrak{K}^R \to \mathfrak{H}\mathfrak{K}(J_1, T) \cong \mathrm{coalg}(T, T)$$

is an epimorphism, but not an isomorphism in general because $\mathfrak{H}\mathfrak{K}^R$ is built up by

$$\widetilde{\mathrm{Lie}}^R(n) = \mathrm{Lie}^{\mathbb{Z}}(n) \otimes_{\mathbb{Z}} R^{\otimes n}$$

according to Theorem 1.4.10 while $\mathrm{coalg}(T,T)$ is built up by $\mathrm{Lie}^R(n)$ by assertions (2) and (3).

Let V be a projective R-module and let $L(V)$ denote the free Lie algebra generated by V. This gives a functor L from projective R-modules to Lie algebras over R. Recall that there is a decomposition

$$L = \bigoplus_{n=1}^{\infty} L_n$$

as functors to R-modules, where $L_n(V)$ is the submodule of $L(V)$ consisting of homogeneous Lie elements of weight n. Let $f \colon T \to T$ be a coalgebra map. Then f maps $L(V)$ into primitive elements $PT(V)$. According to [**39**, Proof of Proposition 2.4], f_V maps $L(V)$ into the submodule $L(V) \subseteq PT(V)$. Moreover the restriction of f_V to $\mathrm{Lie}^R(n) \subseteq L_n(V)$ maps into $\mathrm{Lie}^R(n) \subseteq L_n(V)$. This defines a restriction map

$$\theta \colon \mathrm{coalg}(T,T) \longrightarrow \prod_{n=1}^{\infty} \mathrm{Hom}_{R(S_n)}(\mathrm{Lie}^R(n), \mathrm{Lie}^R(n))$$

with the commutative diagram of semirings

$$\begin{array}{ccc} \mathrm{coalg}(T,T) & \xrightarrow{\theta} & \prod_{k=1}^{\infty} \mathrm{Hom}_{R(S_k)}(\mathrm{Lie}^R(k), \mathrm{Lie}^R(k)) \\ \downarrow & & \downarrow \mathrm{proj.} \\ \mathrm{coalg}(J_n, T) & \xrightarrow{\theta_n} & \prod_{k=1}^{n} \mathrm{Hom}_{R(S_k)}(\mathrm{Lie}^R(k), \mathrm{Lie}^R(k)) \end{array}$$

for each n.

PROPOSITION 2.1.5. *Let ground ring R be any commutative ring with identity.*

1) *The restriction map*

$$\theta_n \colon \text{coalg}(J_n, J_n) = \text{coalg}(J_n, T) \to \prod_{k=1}^{n} \text{Hom}_{R(S_k)}(\text{Lie}^R(k), \text{Lie}^R(k))$$

is a morphism of semirings for each $1 \leq n \leq \infty$, where $J_\infty = T$.

2) *The map θ_n factors through the quotient $\mathfrak{R}(J_n, T)$.*

3) *The resulting map*

$$\bar{\theta}_n \colon \mathfrak{R}(J_n, T) \to \prod_{k=1}^{n} \text{Hom}_{R(S_k)}(\text{Lie}^R(k), \text{Lie}^R(k))$$

is a morphism of rings.

PROOF. The proof of Assertion (1) is immediate and Assertion (3) follows from Assertion (1).

(2). Since $L(V)$ is contained in the set of primitive elements of $T(V)$, the inclusion $L(V) \subseteq T(V)$ induces a morphism of coalgebras

$$j_n \colon J_1\left(\bigoplus_{k=1}^{n} L_k(V)\right) \longrightarrow J_n(V)$$

and so there is a commutative diagram

$$\begin{array}{ccc}
\text{coalg}(J_n, T) & \xrightarrow{j_n^*} & \text{coalg}(J_1 \circ (\oplus_{k=1}^n L_k), T) \\
\downarrow & & \downarrow \\
\mathfrak{R}(J_n, T) & \longrightarrow & \mathfrak{R}(J_1 \circ (\oplus_{k=1}^n L_k), T).
\end{array}$$

Since $J_1\left(\bigoplus_{k=1}^{n} L_k(V)\right)$ has the trivial comultiplication, the map

$$\text{coalg}(J_1 \circ (\oplus_{k=1}^n L_k), T) \longrightarrow \mathfrak{R}(J_1 \circ (\oplus_{k=1}^n L_k), T)$$

is an isomorphism. Note that θ_n factors through j_n^*. The assertion follows. □

Let $\mathfrak{I}_n^R$ denote the kernel of $\mathfrak{R}(J_n, T) \to \mathfrak{R}(J_{n-1}, T)$. Then there is commutative diagram

$$\begin{array}{ccc}
\mathfrak{I}_n^R \hookrightarrow & \mathfrak{R}(J_n, T) & \twoheadrightarrow \mathfrak{R}(J_{n-1}, T) \\
\downarrow \bar{\theta}_n & \downarrow \bar{\theta}_n & \downarrow \bar{\theta}_{n-1} \\
\text{End}_{R(S_n)}(\text{Lie}^R(n)) \hookrightarrow & \prod_{k=1}^{n} \text{End}_{R(S_k)}(\text{Lie}^R(k)) & \twoheadrightarrow \prod_{k=1}^{n-1} \text{End}_{R(S_k)}(\text{Lie}^R(k)).
\end{array}$$

Some terminologies on the stable category of modules over an algebra (symmetric group algebra $R(S_n)$ in our case) will be used in the proof of the following theorem. (See for instance [**25**, Chapter 14].) Let $\bar{V}$ be a free R-module of rank n with a given basis $\{x_1, \ldots, x_n\}$. Let γ_n be the submodule of $\bar{V}^{\otimes n}$ spanned by

$$x_{\sigma(1)} x_{\sigma(2)} \cdots x_{\sigma(n)}$$

for $\sigma \in S_n$. Let S_n act on γ_n^R by permuting letters. Note that $\gamma_n \cong R(S_n)$ is a free $R(S_n)$-module and $\mathrm{Lie}^R(n)$ is a $R(S_n)$-submodule of γ_n. Let M and N be $R(S_n)$-modules and let $f, g\colon M \to N$ be $R(S_n)$-maps. The map f is called to be *homotopic* to g, denoted by $f \simeq g$, if the difference $f - g\colon M \to N$ factors through a projective $R(S_n)$-module. Let

$$[M, N] = \mathrm{Hom}_{R(S_n)}(M, N)/\simeq$$

denote the set of homotopy classes from M to N. Note that $[M, N] = H_0^{R(S_n)}(M, N)$.

THEOREM 2.1.6. *For each $n \geq 1$, there is an exact sequence of abelian groups*

$$0 \longrightarrow \mathrm{Ext}_{R(S_n)}(\mathrm{Lie}^R(n)^*, \mathrm{Lie}^R(n)^*) \longrightarrow \mathfrak{I}_n \xrightarrow{\bar{\theta}_n} \mathrm{Hom}_{R(S_n)}(\mathrm{Lie}^R(n), \mathrm{Lie}^R(n))$$

with the cokernel of $\bar{\theta}_n$ given by $H_{R(S_n)}^0(\mathrm{Lie}^R(n)^*, \mathrm{Lie}^R(n)^*)$.

PROOF. Let $T \wedge T$ be filtered by product filtration. Then there is a commutative diagram of short exact sequences of functors from projective R-modules to coalgebras

$$\begin{array}{ccccc}
J_{n-1} & \hookrightarrow & J_n & \twoheadrightarrow & J_n/J_{n-1} = J_1 \circ T_n \\
\downarrow \bar{\psi} & & \downarrow \bar{\psi} & & \downarrow \mathrm{sh} \\
\mathrm{Fil}_{n-1}(T \wedge T) & \hookrightarrow & \mathrm{Fil}_n(T \wedge T) & \twoheadrightarrow & J_1 \circ (\oplus_{i=1}^{n-1} T_i \otimes T_{n-i}),
\end{array}$$

where the composite

$$T_n(V) = V^{\otimes n} \xrightarrow{\mathrm{sh}} \bigoplus_{i=1}^{n-1} T_i(V) \otimes T_{n-i}(V) \xrightarrow{\mathrm{proj.}} T_i(V) \otimes T_{n-i}(V) = V^{\otimes n}$$

is the shuffle map of type $(i, n-i)$. The following commutative diagram of groups plays an important role:
(2.1.1)

$$\begin{array}{ccccc}
\mathrm{coalg}(J_1 \circ (\bigoplus_{i=1}^{n-1} T_i \otimes T_{n-i}), T) & \hookrightarrow & \mathrm{coalg}(\mathrm{Fil}_n(T^{\wedge 2}), T) & \xrightarrow{i_n^*} & \mathrm{coalg}(\mathrm{Fil}_{n-1}(T^{\wedge 2}), T) \\
\downarrow \mathrm{sh}^* & & \downarrow \bar{\psi}^* & & \downarrow \bar{\psi}^* \\
\mathrm{coalg}(J_1 \circ T_n, T) & \hookrightarrow & \mathrm{coalg}(J_n, T) & \twoheadrightarrow & \mathrm{coalg}(J_{n-1}, T) \\
\downarrow & & \downarrow & & \downarrow \\
\mathfrak{I}_n^R & \hookrightarrow & \mathfrak{R}(J_n, T) & \twoheadrightarrow & \mathfrak{R}(J_{n-1}, T).
\end{array}$$

By proposition 2.1.2 and Theorem 2.1.3, the right two columns are right short exact and the middle row is short exact. For checking the short exactness of the top row, it suffices to show that the map

$$i_n^*\colon \mathrm{coalg}(\mathrm{Fil}_n(T \wedge T), T) \longrightarrow \mathrm{coalg}(\mathrm{Fil}_{n-1}(T \wedge T), T)$$

is onto. Let JC be the free Hopf extension of the functor C to coalgebras, that is, JC is the tensor algebra generated by IC with the (unique) coalgebra structure induced

2.1. FUNCTORS TO COALGEBRAS

from C. Following from the lines in [40, Section 3.1], there is an isomorphism of functors to Hopf algebras

$$(2.1.2) \qquad J\operatorname{Fil}_n(T \wedge T) \cong J\left(\bigvee_{i+j \leq n} (J_1 \circ T_i) \wedge (J_1 \circ T_j)\right).$$

(See the remark below for a detailed proof for this decomposition formula.) For any coalgebra map $f\colon \operatorname{Fil}_{n-1}(T \wedge T) \to T$, let $Jf\colon J\operatorname{Fil}_{n-1}(T \wedge T) \to T$ be the unique Hopf map such that $Jf|_{\operatorname{Fil}_{n-1}(T \wedge T)} = f$. Then the composite

$$\operatorname{Fil}_n(T \wedge T) \hookrightarrow J\operatorname{Fil}_n(T \wedge T) \cong J\left(\bigvee_{i+j \leq n} (J_1 \circ T_i) \wedge (J_1 \circ T_j)\right)$$

$$\xrightarrow{\text{proj}} J\left(\bigvee_{i+j \leq n-1} (J_1 \circ T_i) \wedge (J_1 \circ T_j)\right) \cong J\operatorname{Fil}_{n-1}(T \wedge T) \xrightarrow{Jf} T$$

is a coalgebra extension of f. Thus i_n^* is onto and so the top row is short exact. It follows that $\mathfrak{J}_n^R$ is the cokernel of the map

$$\operatorname{sh}^*\colon \bigoplus_{i=1}^{n-1} \operatorname{Lie}^R(n) = \operatorname{coalg}(J_1 \circ (\oplus_{i=1}^{n-1} T_i \otimes T_{n-i}), T)$$

$$\longrightarrow \operatorname{Lie}^R(n) = \operatorname{coalg}(J_1 \circ T_n, T) = \operatorname{Hom}_R(T_n, L_n),$$

where $\operatorname{Lie}^R(n) = \operatorname{coalg}(J_1 \circ T_n, T)$ by assertion (2) of Theorem 2.1.3 and

$$\operatorname{Lie}^R(n) = \operatorname{Hom}_R(T_n, L_n)$$

follows the lines in [39, Sections 2-3].

Now we start to determine the cokernel of $\bar{\theta}_n$. The plan for identifying the cokernel of $\bar{\theta}_n$ is to pre-compose with $\operatorname{coalg}(J_1 \circ T_n, T) \to \mathfrak{J}_n^R$ and consider the cokernel of θ_n. Let $\alpha \in \operatorname{Hom}_R(T_n, L_n)$. The functorial map

$$\alpha_V\colon T_n(V) = V^{\otimes n} \to L_n(V)$$

induces an $R(S_n)$-map $\alpha\colon \gamma_n \to \operatorname{Lie}^R(n)$. Note that the restriction

$$\theta_n(\alpha)\colon \operatorname{Lie}^R(n) \to \operatorname{Lie}^R(n)$$

is the composite

$$\theta_n(\alpha)\colon \operatorname{Lie}^R(n) \hookrightarrow \gamma_n \longrightarrow \operatorname{Lie}^R(n).$$

Thus

$$\theta_n(\alpha) \simeq 0\colon \operatorname{Lie}^R(n) \to \operatorname{Lie}^R(n)$$

for any $\alpha \in \operatorname{Hom}_R(T_n, L_n)$. Let $f\colon \operatorname{Lie}^R(n) \to \operatorname{Lie}^R(n)$ be an $R(S_n)$-map such that $f \simeq 0$. Then there exists an $R(S_n)$-map $g\colon \gamma_n \to \operatorname{Lie}^R(n)$ such that f is the restriction of g to $\operatorname{Lie}^R(n)$ by using that property that $R(S_n)$ is self-dual. Let $\alpha'\colon T_n \to L_n(V)$ be the natural transformation corresponding to g. Then, since f is the restriction of g to $\operatorname{Lie}^R(n)$,

$$\theta_n(\alpha') = f.$$

It follows that the cokernel of

$$\theta_n\colon \operatorname{Hom}_R(T_n, L_n) \longrightarrow \operatorname{Hom}_{R(S_n)}(\operatorname{Lie}^R(n), \operatorname{Lie}^R(n))$$

is isomorphic to $[\mathrm{Lie}^R(n), \mathrm{Lie}^R(n)]$. Since the map
$$\bar{\theta}_n \colon \mathfrak{I}_n^R \to \mathrm{End}_{R(S_n)}(\mathrm{Lie}^R(n))$$
is induced from θ_n, the cokernel of $\bar{\theta}_n$ is isomorphic to
$$[\mathrm{Lie}^R(n), \mathrm{Lie}^R(n)] \cong H_0^{R(S_n)}(\mathrm{Lie}^R(n), \mathrm{Lie}^R(n)) \cong H^0_{R(S_n)}(\mathrm{Lie}^R(n)^*, \mathrm{Lie}^R(n)^*).$$

Now we show that the kernel of $\bar{\theta}_n$ is isomorphic to $\mathrm{Ext}_{R(S_n)}(\mathrm{Lie}^R(n)^*, \mathrm{Lie}^R(n)^*)$. Consider the commutative diagram of exact sequences

$$\begin{array}{ccccc}
\mathrm{Ker}(\bar{\theta}_n) & \hookrightarrow & \mathfrak{I}_n^R & \xrightarrow{\bar{\theta}_n} & \mathrm{End}_{R(S_n)}(\mathrm{Lie}^R(n)) \\
\uparrow \phi & & \uparrow & & \| \\
\mathrm{Ker}(\theta_n) & \hookrightarrow & \mathrm{Hom}_R(T_n, L_n) & \xrightarrow{\theta_n} & \mathrm{End}_{R(S_n)}(\mathrm{Lie}^R(n)). \\
\uparrow \cong & & \uparrow \cong & & \| \\
\mathrm{Hom}_{R(S_n)}(\bar{\gamma}_n, \mathrm{Lie}^R(n)) & \hookrightarrow & \mathrm{Hom}_{R(S_n)}(\gamma_n, \mathrm{Lie}^R(n)) & \to & \mathrm{End}_{R(S_n)}(\mathrm{Lie}^R(n)),
\end{array}$$

where $\bar{\gamma}_n = \gamma_n / \mathrm{Lie}^R(n)$. Let $\beta \in \mathrm{Ker}(\theta_n)$. Suppose that $\phi(\beta) = 0$. By Diagram (2.1.1), there exits a natural transformation $\tilde{\beta} \colon \bigoplus_{i=1}^{n-1} T_i \otimes T_{n-i} \to L_n$ such that $\beta = \mathrm{sh} \circ \tilde{\beta}$. Consider the diagram

$$\begin{array}{ccccc}
\gamma_n & \xrightarrow{q_n} & \gamma_n / \mathrm{Lie}^R(n) & \xrightarrow{\beta} & \mathrm{Lie}^R(n) \\
\downarrow \mathrm{sh} & & \downarrow & & \| \\
\bigoplus_{i=1}^{n-1} \gamma_n & = & \bigoplus_{i=1}^{n-1} \gamma_n & \xrightarrow{\tilde{\beta}} & \mathrm{Lie}^R(n).
\end{array}$$

Since the outer rectangle commutes, then left square does as $\mathrm{Lie}^R(n) \subseteq \mathrm{Ker}(\mathrm{sh})$ and so the epimorphism q_n implies the right square commutes. Thus the representing $R(S_n)$-map $\beta \colon \gamma_n / \mathrm{Lie}^R(n) \to \mathrm{Lie}^R(n)$ is null homotopic. Conversely, let
$$\delta \colon \gamma_n / \mathrm{Lie}^R(n) \to \mathrm{Lie}^R(n)$$
be a null homotopic $R(S_n)$-map. Then δ factors through $\bigoplus_{i=1}^{n-1} \gamma_n$, that is, there is an $R(S_n)$-map $\tilde{\delta} \colon \bigoplus_{i=1}^{n-1} \gamma_n \to \mathrm{Lie}^R(n)$ such that
$$\tilde{\delta} \circ \mathrm{sh} = \delta \circ q_n \colon \gamma_n \longrightarrow \mathrm{Lie}^R(n).$$

Note that
$$\mathrm{sh} \otimes_{R(S_n)} \mathrm{id}_{V^{\otimes n}} = \bar{\psi}_V \colon \gamma_n \otimes_{R(S_n)} V^{\otimes n} = V^{\otimes n}$$
$$\longrightarrow \bigoplus_{i=1}^{n-1} \gamma_n \otimes_{R(S_n)} V^{\otimes n} = \bigoplus_{i=1}^{n-1} T_i(V) \otimes T_{n-i}(V).$$

The induced natural transformation of $\delta \circ q_n = \tilde{\delta} \circ \text{sh}$ is 0 in $\mathfrak{R}(J_n, T)$. Thus $\phi(\delta \circ q_n) = 0$ in $\mathfrak{I}_n$. This proves that there is an isomorphism

$$\text{Ker}(\bar{\theta}_n) \cong [\gamma_n/\text{Lie}^R(n), \text{Lie}^R(n)].$$

As R-modules, $\text{Lie}^R(n)$ is projective and so is $\gamma_n/\text{Lie}^R(n)$. Thus $\gamma_n/\text{Lie}^R(n)$ is the suspension of $\text{Lie}^R(n)$ and so the dual $(\gamma_n/\text{Lie}^R(n))^*$ is the looping of $\text{Lie}^R(n)^*$. By using the canonical isomorphism

$$\text{Hom}_{R(S_n)}(\gamma_n/\text{Lie}^R(n), \text{Lie}^R(n)) \cong \text{Hom}_{R(S_n)}\left(\text{Lie}^R(n)^*, (\gamma_n/\text{Lie}^R(n))^*\right),$$

there is an isomorphism

$$[\gamma_n/\text{Lie}^R(n), \text{Lie}^R(n)] \cong \text{Ext}_{R(S_n)}(\text{Lie}^R(n)^*, \text{Lie}^R(n)^*)$$

and hence the result. $\square$

REMARK 2.1.7. We give a detailed proof for the decomposition formula (2.1.2) in this remark as it does not seem to be recorded in references. We need a concept for functors to coalgebras:

> A functor C to coalgebras is called *suspension equivalent* to a functor D to coalgebras, denoted by $C \stackrel{s}{\sim} D$, if $JC \cong JD$ as functors to Hopf algebras.

(This is an algebraic analogue of suspension homotopy equivalence in geometry.) By using the algebraic version of the James-Hopf invariants [40, Section 3.1], there is an isomorphism of functors to Hopf algebras

$$JT \cong J\left(\bigvee_{n=1}^{\infty} J_1 \circ T_n\right).$$

Thus there is an algebraic suspension decomposition $T \stackrel{s}{\sim} \bigvee_{n=1}^{\infty} J_1 \circ T_n$ and so

$$T \wedge T \stackrel{s}{\sim} \left(\bigvee_{i=1}^{\infty} J_1 \circ T_j\right) \wedge \left(\bigvee_{i=1}^{\infty} J_1 \circ T_j\right) = \bigvee_{i,j=1}^{\infty} (J_1 \circ T_i) \wedge (J_1 \circ T_j).$$

It follows that

$$\text{Fil}_n(T \wedge T) \stackrel{s}{\sim} \bigvee_{i+j \leq n} J_1 \circ (T_i \wedge T_j),$$

which gives the decomposition formula (2.1.2). $\square$

EXAMPLE 2.1.8. In this example, we are going to display the first obstruction to the exponent problem. The ground ring $R = \mathbb{Z}_{(p)}$ or $\mathbb{Z}/p^r$.

Since $\text{Lie}^R(n)$ is a projective $R(S_n)$-module for $n \not\equiv 0 \mod p$, the extension group

$$\text{Ext}_{R(S_n)}(\text{Lie}^R(n)^*, \text{Lie}^R(n)^*) = 0$$

for $1 < n < p$. It follows that the map

$$\bar{\theta}_{p-1} \colon \mathfrak{R}(J_{p-1}, T) \longrightarrow \prod_{k=1}^{p-1} \text{Hom}_{R(S_k)}(\text{Lie}^R(k), \text{Lie}^R(k))$$

is a monomorphism. In particular, if $R = \mathbb{Z}/p^r$, then $p^r \cdot \mathfrak{R}(J_{p-1}, T) = 0$. In other words, the exponents of $\mathfrak{R}(J_n, T)$ does not increase for $n \leq p - 1$.

We compute $\text{Ext}_{R(S_p)}(\text{Lie}^R(p)^*, \text{Lie}^R(p)^*)$. By [**39**, Corollaries 8.23, 8.27 and 11.10], there is a decomposition of modules over $R(S_p)$

$$\text{Lie}^R(p) \cong \text{Lie}^{\max}(p) \oplus IR^{\oplus p},$$

where $\text{Lie}^{\max}(p)$ is a projective $R(S_p)$-module of dimension $(p-1)! - p + 1$, the symmetric group S_p acts on $R^{\oplus p}$ by the regular representation (that is by permuting the elements in the basis $\{y_1, y_2, \ldots, y_p\}$ for R^p) and $IR^{\oplus p}$ is the kernel of the augmentation map $\epsilon \colon R^{\oplus p} \to R$ with $\epsilon(y_i) = 1$ for $1 \leq i \leq p$. It follows that

$$\text{Ext}_{R(S_p)}(\text{Lie}^R(p)^*, \text{Lie}^R(p)^*) \cong \text{Ext}_{R(S_p)}((IR^{\oplus p})^*, (IR^{\oplus p})^*).$$

From the fact that $R^{\oplus p}$ is a projective $R(S_p)$-module and the short exact sequence

$$IR^{\oplus p} \hookrightarrow R^{\oplus p} \xrightarrow{\epsilon} R,$$

there is an isomorphism

$$\text{Ext}_{R(S_p)}((IR^{\oplus p})^*, (IR^{\oplus p})^*) \cong [R, IR^{\oplus p}] = \text{Hom}_{R(S_p)}(R, IR^{\oplus p})/\simeq$$

the group of the homotopy classes from the trivial module R into $IR^{\oplus p}$. Let $f \colon R \to IR^{\oplus p}$ be a $R(S_p)$-map. Then $\text{Im}(f)$ is the trivial submodule of $IR^{\oplus p}$ generated by $\alpha = f(1)$. Since $IR^{\oplus p}$ has a basis $\{y_2 - y_1, \ldots, y_p - y_1\}$, there exist (unique) $k_2, k_3, \ldots, k_p \in R$ such that

$$\alpha = \sum_{i=2}^{p} k_i(y_i - y_1) = \sum_{i=1}^{p} k_i y_i \in IR^{\oplus p} \subseteq R^{\oplus p}$$

with $k_1 = -k_2 - \cdots - k_p$. Since $\sigma \cdot \alpha = \alpha$ for all $\sigma \in S_p$, we have

$$k_1 = k_2 = \cdots = k_p = k$$

by applying $\tau_{i,j} = (i, j)$ to α. It follows that

$$\alpha = k \sum_{i=2}^{p} (y_i - y_1)$$

with $pk = 0$. Thus

$$\text{Hom}_{R(S_p)}(R, IR^{\oplus p}) = \begin{cases} 0 & \text{if} \quad R = \mathbb{Z}_{(p)}; \\ \mathbb{Z}/p \cong p^{r-1} \cdot R & \text{if} \quad R = \mathbb{Z}/p^r. \end{cases}$$

It follows that the map

$$\bar{\theta}_p \colon \mathfrak{R}(J_p, T) \longrightarrow \prod_{k=1}^{p} \text{Hom}_{R(S_k)}(\text{Lie}^R(k), \text{Lie}^R(k))$$

is a monomorphism for the case when $R = \mathbb{Z}_{(p)}$.

Consider the case when $R = \mathbb{Z}/p^r$. Let $j \colon R \to R^{\oplus p}$ with $j(1) = \sum_{i=1}^{p} y_i$. Then any null-homotopic $R(S_p)$-map $g \colon R \to IR^{\oplus p}$ factors through $j \colon R \to R^{\oplus p}$ because $R^{\oplus p}$ is (smallest) projective $R(S_p)$-module containing R. Let $\tilde{g} \colon R^{\oplus p} \to IR^{\oplus p}$ be an $R(S_p)$-map such that $g = \tilde{g} \circ j$. There exist (unique) $k'_2, \ldots, k'_p \in R$ such that

$$\tilde{f}(y_1) = \sum_{i=2}^{p} k'_i(y_i - y_1) = k'_1 y_1 + \sum_{i=2}^{p} k'_i y_i \in IR^{\oplus p} \subseteq R^{\oplus p},$$

where $k_1' = -k_2' - \cdots - k_p'$. By applying $\tau_{i,j} = (i,j)$ for $i,j > 1$, we have

$$\tau_{i,j} \cdot \sum_{i=2}^{p} k_i' y_i = \sum_{i=2}^{p} k_i' y_i$$

and so $k_2' = k_3' = \cdots = k_p' = k$ for some $k \in R$. Thus

$$\tilde{g}(y_1) = k \sum_{i=2}^{p} (y_i - y_1) = k \sum_{i=1}^{p} y_i - pky_1$$

and so

$$\tilde{g}(y_i) = k \sum_{i=1}^{p} y_i - pky_i$$

for $1 \leq i \leq p$. It follows that

$$g(1) = \tilde{g} \circ j(1) = \sum_{i=1}^{p} \tilde{f}(y_i) = k \sum_{i=1}^{p} \left(\sum_{i=1}^{p} y_i - py_i \right) = 0$$

and so

$$\mathrm{Ext}_{R(S_p)}((IR^{\oplus p})^*, (IR^{\oplus p})^*) \cong [R, IR^{\oplus p}] \cong \mathrm{Hom}_{R(S_p)}(R, IR^{\oplus p}) \cong \mathbb{Z}/p$$

for $R = \mathbb{Z}/p^r$ with a generator represented by the map

$$\zeta \colon R \to IR^{\oplus p} \quad 1 \mapsto p^{r-1} \cdot \sum_{i=2}^{p} (y_i - y_1) = p^{r-1} \sum_{i=1}^{p} y_i.$$

Thus there is a left short exact sequence

$$0 \longrightarrow \mathbb{Z}/p \longrightarrow \mathfrak{R}(J_p, T) \longrightarrow \prod_{k=1}^{p} \mathrm{Hom}_{R(S_k)}(\mathrm{Lie}^R(k), \mathrm{Lie}^R(k))$$

for $R = \mathbb{Z}/p^r$. It follows that $p^{r+1} \cdot \mathfrak{R}(J_p, T) = 0$, that is $\mathfrak{R}(J_p, T)$ has an exponent at most p^{r+1}.

For showing that the exponent of $\mathfrak{R}(J_p, T)$ is exactly equal to p^{r+1}, consider the element $p^r \cdot \mathrm{id}_T |_{J_p}$ in $\mathfrak{R}(J_p, T)$ that is represented, as a functorial self coalgebra map of tensor algebras, by the composite

$$\mathrm{id}_T^{*p^r} \colon T \xrightarrow{\psi^{p^r-1}} T^{\otimes p^r} \xrightarrow{\text{multi.}} T$$

the p^r-th convolution power of the identity id_T. Note that $\mathrm{Hom}_R(T,T)$ is an algebra under convolution product with $\mathrm{coalg}(T,T) \subseteq \mathrm{Hom}_R(T,T)$. The identity 1 of the ring $\mathrm{Hom}_R(T,T)$ is the composite $T \xrightarrow{\epsilon} R \xrightarrow{\eta} T$. Let $q = \mathrm{id}_T - 1$ in $\mathrm{Hom}_R(T,T)$ which is represented by the composite $T \xrightarrow{\text{proj.}} IT \hookrightarrow T$. Then

$$\mathrm{id}_T^{*p^r} = (1+q)^{p^r} = 1 + \sum_{i=1}^{p^r} \binom{p^r}{i} q^i = 1 + \binom{p^r}{p} q^p + \sum_{i=p+1}^{p^r} \binom{p^r}{i} q^i$$

in $\mathrm{Hom}_R(T,T)$ because $\binom{p^r}{i} \equiv 0 \mod p^r$ for $1 < i < p$. Since q^i is represented by the composite

$$T \xrightarrow{\psi^{i-1}} T^{\otimes i} \xrightarrow{\text{proj.}} IT^{\otimes i} \xrightarrow{\text{multi.}} T,$$

the restriction map $q^i|_{J_p} = 0$ for $i > p$ because $IT^{\otimes i}$ is a summation of T_i and a collection of T_j's with $j > i$. Moreover $q^p|_{J_p}$ is represented by the composite

$$J_p(V) \xrightarrow{\text{proj.}} T_p(V) = V^{\otimes p} \xrightarrow{\sum_{\sigma \in S_p} \sigma} V^{\otimes p} \hookrightarrow T(V),$$

where S_p acts on $V^{\otimes p}$ by permuting positions. It follows that

$$\text{id}_T^{*p^r} \in \text{Ker}(\text{coalg}(J_p, T) \to \text{coalg}(J_{p-1}, T)) \cong \text{Lie}^R(p)$$

represented by the element $\binom{p^r}{p} \sum_{\sigma \in S_p} \sigma$. We need to rewrite this element in terms of Lie elements. Let $\bar{V}_n$ is the free R-module with a basis $\{x_1, \ldots, x_n\}$ and let

$$\text{tr}_n = \sum_{\sigma \in S_n} x_{\sigma(1)} x_{\sigma(2)} \cdots x_{\sigma(n)} \in \gamma_n \subseteq \bar{V}^{\otimes n} \quad \text{and}$$

$$\overline{\text{tr}}_n = \sum_{\tau \in S_{n-1}} [[x_1, x_{\tau(2)}], x_{\tau(3)}, \ldots, x_{\tau(n)}] \in \text{Lie}^R(n)$$

be the summations of the standard basis for γ_n and $\text{Lie}^R(n)$, respectively. Then, for $R = \mathbb{Z}/p^r$,

$$p^{r-1}(x_1 + x_2 + \cdots + x_{p^t})^{p^t} = p^{r-1} \text{tr}_{p^t} + W$$

$$= p^{r-1}(x_1 + v)^{p^t} = p^{r-1}[[x_1, v], v, \ldots, v] + W' = p^{r-1} \overline{\text{tr}}_{p^t} + W,$$

where $v = x_2 + x_3 + \cdots + x_{p^t}$, W is a sum of the homogenous terms in which one of x_i occurs at least twice and W' is a sum of the homogenous terms in which the occurrence of x_1 is 0 or ≥ 2. Thus

$$p^{r-1} \text{tr}_{p^t} = p^{r-1} \overline{\text{tr}}_{p^t}$$

and so

$$\text{id}_T^{*p^r} = \binom{p^r}{p} \overline{\text{tr}}_p$$

in $\text{Lie}^R(p)$. According to [**39**, Proposition 11.1], the projection $\gamma_p \to R^{\oplus p}$ is given by sending the monomials $x_{\sigma(1)} x_{\sigma(2)} \cdots x_{\sigma(p)}$ to $y_{\sigma(p)}$ for $\sigma \in S_p$. Hence $\text{id}_T^{*p^r}$ is represented by

$$\binom{p^r}{p} \cdot (p-1)! \sum_{i=1}^p y_p = \frac{(p^r - 1)!}{(p^r - p)!} \cdot p^{r-1} \sum_{i=2}^p (y_i - y_1) = \frac{(p^r - 1)!}{(p^r - p)!} \zeta(1)$$

in $IR^{\oplus p}$, which is a generator for $\text{Ext}_{R(S_p)}(\text{Lie}^R(p)^*, \text{Lie}^R(p)^*)$. This finishes our computation for showing that $\exp(\mathfrak{R}(J_p, T)) = p^{r+1}$. $\square$

Remark. Following from the lines in Example 2.1.8, for $R = \mathbb{Z}/p^r$, the convolution power

$$\text{id}_T^{*p^{r+t}} \in \text{Ker}(\text{coalg}(J_{p^{t+1}}, T) \to \text{coalg}(J_{p^{t+1}-1}, T)) = \text{Lie}^R(p^{t+1})$$

represented by $\binom{p^{r+t}}{p^{t+1}} \cdot \overline{\text{tr}}_{p^{t+1}}$. It is unknown whether this obstruction vanishes in the extension group $\text{Ext}_{R(S_{p^{t+1}})}(\text{Lie}^R(p^{t+1})^*, \text{Lie}^R(p^{t+1})^*)$ or not for general $t > 0$. In geometry, by applying the Cohen group, the element $\overline{\text{tr}}_n$ is represented by the composite

$$\overline{\text{tr}}_n : J_n(\Sigma X) \xrightarrow{\text{pinch}} (\Sigma X)^{(n)} \xrightarrow{\sum_{\tau \in S_{n-1}} \text{id}_{\Sigma X} \wedge \tau} (\Sigma X)^{(n)} \xrightarrow{W_n} \Omega \Sigma^2 X \xrightarrow{\Omega f} \Omega Y$$

for $f\colon \Sigma^2 X \to Y$. Thus, if $p^r[f] = 0$ in $[\Sigma^2 X, Y]$, then the power $(\Omega f)^{p^{r+t}}$ is null homotopic restricted to $J_{p^{t+1}-1}(\Sigma X)$ and homotopic to $\binom{p^{r+t}}{p^{t+1}} \cdot \overline{\mathrm{tr}}_{p^{t+1}}$ restricted to $J_{p^{t+1}}(\Sigma X)$. In the case when f is the identity of a Moore space $P^n(p^r)$ with $p > 2$, then $\Omega^2(f)^{p^{r+1}}$ is null homotopic according to the classical exponent results in [**31**] and so the obstruction $\binom{p^{r+t}}{p^{t+1}} \cdot \overline{\mathrm{tr}}_{p^{t+1}}$ vanishes after looping for $t > 0$ for these particular spaces.

2.2. Geometric Realizations and the Shuffle Relations

Let $\mathfrak{K}(X, \Omega Y) = \{[X^{n+1}, \Omega Y]\}_{n \geq 0}$ be the bi-Δ-group with faces and cofaces described in Subsection 1.2.

LEMMA 2.2.1. *Let*
$$q\colon X^{l_1+l_2+\cdots+l_t} \longrightarrow J_{l_1}(X) \times J_{l_2}(X) \times \cdots \times J_{l_t}(X)$$
be the quotient map. Then
$$q^*\colon [J_{l_1}(X) \times J_{l_2}(X) \times \cdots \times J_{l_t}(X), \Omega Y] \longrightarrow [X^{l_1+l_2+\cdots+l_t}, \Omega Y]$$
is a monomorphism with the image given by the elements $\alpha \in [X^{l_1+l_2+\cdots+l_t}, \Omega Y]$ such that
$$d_j \alpha = d_{l_1+l_2+\cdots+l_s-1} \alpha$$
for $l_1 + l_2 + \cdots + l_{s-1} \leq j \leq l_1 + l_2 + \cdots + l_s - 1$ and $1 \leq s \leq t$ in $\mathfrak{K}(X, \Omega Y)$.

PROOF. By the suspension splitting theorem, q^* is a monomorphism. The second part follows from Proposition 1.1.1 by induction on t. □

COROLLARY 2.2.2. *Let $q\colon X^{s+t+1} = X^{s+1+t} \to J_{s+1}(X) \wedge J_t(X)$ be the quotient map. Then*
$$q^*\colon [J_{s+1}(X) \wedge J_t(X), \Omega Y] \longrightarrow [X^{s+t+1}, \Omega Y]$$
is a monomorphism with the image consisting of the elements
$$\alpha \in \mathfrak{K}(X, \Omega Y)_{s+t} = [X^{s+t+1}, \Omega Y]$$
satisfying the following equations

(2.2.1)
$$\begin{cases} d_0 \alpha = d_1 \alpha = \cdots = d_s \alpha \\ d_{s+1} \alpha = d_{s+2} \alpha = \cdots = d_{s+t} \alpha \\ d_0 d_1 \cdots d_s \alpha = 1 \\ d_{s+1} d_{s+2} \cdots d_{s+t} \alpha = 1, \end{cases}$$

where $d_\emptyset = \mathrm{id}$ in case $t = 0$. □

The diagonal map $\Delta_n\colon X^{n+1} \to X^{n+1} \times X^{n+1}$ induces a group homomorphism
$$\Delta_n^*\colon \mathfrak{K}_{2n+1}(X, \Omega Y) \to \mathfrak{K}_n(X, \Omega Y).$$
It is routine to check that the following identities hold in $\mathfrak{K}(X, \Omega Y)$:
$$d^i \circ \Delta_{n-1}^* = \Delta_n^* d^{n+i+1} d^i$$
$$d_i \Delta_n^* = \Delta_{n-1}^* d_i d_{n+i+1}$$
for $0 \leq i \leq n$.

The proof of the following proposition is routine.

PROPOSITION 2.2.3. *Let $\mathcal{G}$ be a Δ-group and let $\Phi_n \colon \mathcal{G}_{2n+1} \to \mathcal{G}_n$ be a sequence of functions such that $d_i \Phi_n = \Phi_{n-1} d_i d_{n+i+1}$ for $0 \leq i \leq n$. Let $\bar{\mathfrak{H}}_{s,t}$ denote the subgroup of $\mathcal{G}_{s+t}$ consisting of elements that satisfy the equations (2.2.1). Then*

1) *d_i maps $\bar{\mathfrak{H}}_{s,t}\mathcal{G}$ into $\bar{\mathfrak{H}}_{s-1,t}\mathcal{G}$ for $0 \leq i \leq s$.*
2) *d_i maps $\bar{\mathfrak{H}}_{s,t}\mathcal{G}$ into $\bar{\mathfrak{H}}_{s,t-1}$ for $i > s$.*
3) *Φ_n maps $\bar{\mathfrak{H}}_{n,n+1}\mathcal{G}$ into $\mathfrak{H}_n\mathcal{G}$ for each $n \geq 0$.* □

Let G be a group and let $\phi \colon G \to [X, \Omega Y]$ be a group homomorphism. Let

$$\Phi_n \colon \mathfrak{F}^G_{2n+1} = \coprod_{j=0}^{2n+1} (G)_{x_j} \to \mathfrak{F}^G_n = \coprod_{j=0}^{n} (G)_{x_j}$$

be the (unique) group homomorphism such that

$$\Phi_n((g)_{x_j}) = \begin{cases} (g)_{x_j} & \text{if} \quad 0 \leq j \leq n \\ (g)_{x_{j-n-1}} & \text{if} \quad n+1 \leq j \leq 2n+1 \end{cases}$$

for $g \in G$. Then there is a commutative diagram

$$\begin{array}{ccc} \mathfrak{F}^G_{2n+1} & \xrightarrow{e_\phi} & \mathfrak{K}(X, \Omega Y)_{2n+1} = [X^{2n+2}, \Omega Y] \\ \Big\downarrow \Phi_n & & \Big\downarrow \Delta_n^* \\ \mathfrak{F}^G_n & \xrightarrow{e_\phi} & \mathfrak{K}(X, \Omega Y)_n = [X^{n+1}, \Omega Y] \end{array}$$

for each $n \geq 0$. From Relation (1.4.2), the composite

$$\mathfrak{F}^G_{2n+1} \xrightarrow{\Phi_n} \mathfrak{F}^G_n \longrightarrow \mathfrak{K}^{G,k}_n$$

factors through the quotient $\mathfrak{K}^{G,k}_{2n+1}$. The resulting group homomorphism

$$\Phi_n \colon \mathfrak{K}^{G,k}_{2n+1} \to \mathfrak{K}^{G,k}_n$$

has the property that

$$\Delta_n^* \circ e_\phi = e_\phi \circ \Phi_n$$

if the representation ϕ has weak *LS*-category less than or equal to k. This gives the follows proposition.

PROPOSITION 2.2.4. *Let G be a group and let $\phi \colon G \to [X, \Omega Y]$ be a group homomorphism of weak LS-category less than or equal to k. Then there is a commutative diagram*

$$\begin{array}{ccc} \bar{\mathfrak{H}}_{n,n+1}\mathfrak{K}^{G,k} & \xrightarrow{e_\phi} & \bar{\mathfrak{H}}_{n,n+1}\mathfrak{K}(X, \Omega Y) = [J_{n+1}(X) \wedge J_{n+1}(X), \Omega Y] \\ \Big\downarrow \Phi_n & & \Big\downarrow \bar{\Delta}^* \\ \mathfrak{H}_n \mathfrak{K}^{G,k} & \xrightarrow{e_\phi} & \mathfrak{H}_n \mathfrak{K}(X, \Omega Y) = [J_{n+1}(X), \Omega Y] \end{array}$$

for each $n \geq 0$. □

Let $\mathfrak{R}^{G,k}_n$ be the cokernel of the homomorphism $\Phi_n \colon \bar{\mathfrak{H}}_{n,n+1}\mathfrak{K}^{G,k} \longrightarrow \mathfrak{H}_n \mathfrak{K}^{G,k}$.

2.2. GEOMETRIC REALIZATIONS AND THE SHUFFLE RELATIONS

COROLLARY 2.2.5. *Let G be a group and let $\phi\colon G \to [X, \Omega Y]$ be a group homomorphism of weak LS-category less than or equal to k. Then the composite*

$$\mathfrak{H}_n\mathfrak{K}^{G,k} \xrightarrow{e_\phi} \mathfrak{H}_n\mathfrak{K}(X, \Omega Y) = [J_{n+1}(X), \Omega Y] \xrightarrow{\Omega} [\Omega J_{n+1}(X), \Omega^2 Y]$$

factors through $\mathfrak{R}_n^{G,k}$.

PROOF. The assertion follows from the fact that

$$\Omega\bar{\Delta}\colon \Omega(J_{n+1}(X) \wedge J_{n+1}(X)) \to \Omega J_{n+1}(X)$$

is null homotopic. □

THEOREM 2.2.6 (Geometric Realization Theorem). *Let R be a subquotient ring of $\mathbb{Q}$. Then*

$$\mathfrak{R}_n^{R,1} \cong \mathfrak{R}(J_{n+1}, T)$$

for $0 \leq n \leq \infty$, where $J_n = \bigoplus_{i=0}^{n} T_i$ is a functor from projective R-modules to coalgebras over R.

PROOF. Following from the lines in [**39**, Section 2], the group $\mathrm{coalg}(J_1^{\otimes n+1}, T)$ is isomorphic to $\mathfrak{K}_n^{R,1}$ with generators given by the composites

$$x_i\colon J_1^{\otimes n+1} = J_1^{\otimes i} \otimes J_1 \otimes J_1^{\otimes n-i} \xrightarrow{\epsilon_{J_1^{\otimes i}} \otimes \mathrm{id}_{J_1} \otimes \epsilon_{J_1^{\otimes n-i}}} J_1 \hookrightarrow T$$

for $0 \leq i \leq n$. Let $\psi\colon J_1^{\otimes n+1} \longrightarrow J_1^{\otimes n+1} \otimes J_1^{\otimes n+1} = J_1^{\otimes 2n+2}$ be the comultiplication. Then there is a commutative diagram

$$\begin{array}{ccc} \mathfrak{K}_n^{R,1} & \xrightarrow{\cong} & \mathrm{coalg}(J_1^{\otimes n+1}, T) \\ \Phi_n \uparrow & & \uparrow \psi^* \\ \mathfrak{K}_{2n+1}^{R,1} & \xrightarrow{\cong} & \mathrm{coalg}(J_1^{\otimes 2n+2}, T). \end{array}$$

Note that the sequence of groups $\{\mathrm{coalg}(J_1^{\otimes n+1}, T)\}_{n\geq 0}$ is a bi-Δ-group with

$$\mathfrak{H}_n\{\mathrm{coalg}(J_1^{\otimes n+1}, T)\}_{n\geq 0} = \mathrm{coalg}(J_{n+1}, T),$$

$$\bar{\mathfrak{H}}_{n,n+1}\{\mathrm{coalg}(J_1^{\otimes n+1}, T)\}_{n\geq 0} = \mathrm{coalg}(J_{n+1} \wedge J_{n+1}, T).$$

There is a commutative diagram

$$\begin{array}{ccc} \mathfrak{H}_n\mathfrak{K}^{R,1} & \xrightarrow{\cong} & \mathrm{coalg}(J_{n+1}, T) \\ \Phi_n \uparrow & & \uparrow \bar{\psi}^* \\ \bar{\mathfrak{H}}_{n,n+1}\mathfrak{K}^{R,1} & \xrightarrow{\cong} & \mathrm{coalg}(J_{n+1} \wedge J_{n+1}, T). \end{array}$$

Thus $\mathfrak{R}_n^{R,1}$ is isomorphic to $\mathrm{Coker}\left(\bar{\psi}^*\colon \mathrm{coalg}(J_{n+1} \wedge J_{n+1}, T) \to \mathrm{coalg}(J_{n+1}, T)\right)$ and the assertion follows from Part (1) of Proposition 2.1.2. □

2.3. Proof of Theorem 3

Let the ground ring R be any commutative ring with identity. Write $\mathfrak{R}_n^R$ for $\mathfrak{R}(J_n, T)$. Then $\mathfrak{R}^R = \mathfrak{R}(T,T)$ is given by the inverse limit

$$\mathfrak{R}^R = \lim_n \mathfrak{R}_n^R.$$

PROOF OF THEOREM 3. Assertion (1) follows from the fact that

$$\mathfrak{R}^R = \mathfrak{R}(T,T) \cong H_1(\mathcal{W}(T,T); \mathbb{Z})$$

by Proposition 2.1.2. Assertion (2) is obvious. The first part of Assertion (3) is part (3) of Proposition 2.1.5. The second part of Assertion (3) follows from Theorem 2.1.6 by using the fact that $\mathrm{Lie}^R(n)$ is a projective $R(S_n)$-module if R is of characteristic 0. Assertion (4) follows from Remark 2.1.4 and part (1) of Proposition 2.1.5. Assertion (5) is Theorem 2.1.6.

(6). Let $R = \mathbb{Z}_{(p)}$ or $\mathbb{Z}_p$. Consider the diagram

$$\begin{array}{ccccccc}
\mathfrak{H}^R & \longrightarrow & \mathfrak{R}^R & = & \mathfrak{R}^R & \longrightarrow & \mathfrak{R}^{\mathbb{Z}/p^r} \\
\downarrow e & & \downarrow e & & \downarrow e_X & & \downarrow \\
[\Omega\Sigma^2, \Omega\Sigma^2] & \xrightarrow{\Omega} & [\Omega^2\Sigma^2, \Omega^2\Sigma^2] & \longrightarrow & [\Omega^2\Sigma^2 X, \Omega^2\Sigma^2 X] & \xrightarrow{\Omega^2 f} & [\Omega^2\Sigma^2 X, \Omega^2 Y]
\end{array}$$

for $f : \Sigma^2 X \to Y$ of order p^r in $[\Sigma^2 X, Y]$, where $\mathfrak{R}^{\mathbb{Z}/p^r} = \mathfrak{R}(T,T)$ over the ground ring $R = \mathbb{Z}/r$. The left two squares commute by Corollary 2.2.5 and Theorem 2.2.6. By Corollary 1.4.4, the composite

$$\mathfrak{F}^{\mathbb{Z}/} \xrightarrow{e} \mathfrak{K}(\Sigma X, \Omega\Sigma^2 X) \xrightarrow{\Omega f_*} \mathfrak{K}(\Sigma X, \Omega Y)$$

factors through the bi-Δ quotient group $\mathfrak{K}^{\mathbb{Z}/p^r}$ because

$$\Omega f_* \colon \mathfrak{K}(\Sigma X, \Omega\Sigma^2 X)_0 = [\Sigma X, \Omega\Sigma^2 X] \longrightarrow \mathfrak{K}(\Sigma X, \Omega Y)_0 = [\Sigma X, \Omega Y]$$

is of order p^r. Then the rightmost square commutes by Corollary 2.2.5 and Theorem 2.2.6. It suffices to show that there is a commutative diagram

(2.3.1)
$$\begin{array}{ccc}
\mathfrak{R}^R & \xrightarrow{\theta} & \prod_{n=1}^{\infty} \mathrm{Hom}_{R(S_n)}(\mathrm{Lie}^R(n), \mathrm{Lie}^R(n)) \\
\downarrow e & & \uparrow \lambda \\
[\Omega^2\Sigma^2, \Omega^2\Sigma^2] & \xrightarrow{\Omega^{k-2}} & [\Omega^k\Sigma^2, \Omega^k\Sigma^2]
\end{array}$$

for some morphism of semirings λ.

Let CW denote the category of pointed CW-complexes. Let $\mathcal{C}$ be the subcategory of CW with objects given by finite self wedges of S^{2k-1}. The inclusion $\mathcal{C} \subseteq \mathrm{CW}$ induces a morphism of semirings

$$\lambda_1 \colon [\Omega^k\Sigma^2, \Omega^k\Sigma^2] \longrightarrow [\Omega^k\Sigma^2, \Omega^k\Sigma^2]_{\mathcal{C}},$$

where $[\Omega^k\Sigma^2, \Omega^k\Sigma^2]_{\mathcal{C}}$ is the set of natural transformations of the functor $\Omega^k\Sigma^2$ from $\mathcal{C}$ to the homotopy category of spaces, with canonical semiring structure. For each

$X \in \mathcal{C}$ and $f \in [\Omega^k\Sigma^2, \Omega^k\Sigma^2]_\mathcal{C}$, there is an algebraic map

$$(\Omega f_X)_*\colon H_*(\Omega^{k+1}\Sigma^2 X; R) \longrightarrow H_*(\Omega^{k+1}\Sigma^2 X; R)$$

and so a morphism of R-modules

$$Q((\Omega f_X)_*)\colon Q(H_*(\Omega^{k+1}\Sigma^2 X; R))/\mathrm{Tor} \longrightarrow Q(H_*(\Omega^{k+1}\Sigma^2 X; R))/\mathrm{Tor},$$

where $Q(A)$ is the set of indecomposable elements of a Hopf algebra A and $Q(A)/\mathrm{Tor}$ is the quotient group $Q(A)$ by its torsion subgroup. This defines a morphism of semirings

$$\lambda_2 \colon [\Omega^k\Sigma^2, \Omega^k\Sigma^2]_\mathcal{C} \longrightarrow \mathrm{End}_\mathcal{C}(Q(H_*(\Omega^{k+1}\Sigma^2(-); R))/\mathrm{Tor}).$$

Let $\sigma\colon \Sigma^k\Omega^{k+1}\Sigma^2 Z \to \Omega\Sigma^2 Z$ be the evaluation map. Then there is an isomorphism

$$\sigma_*\colon \Sigma^k Q(H_*(\Omega^{k+1}\Sigma^2 X; R))/\mathrm{Tor} \xrightarrow{\cong} P(H_*(\Omega\Sigma^2 X; R)) = L(\bar{H}_*(\Sigma X; R))$$

for $X \in \mathcal{C}$ by considering the Milnor-Hilton Theorem [**20, 29**]. Thus there is an isomorphism of rings

$$\lambda_3 \colon \mathrm{End}_\mathcal{C}(Q(H_*(\Omega^{k+1}\Sigma^2(-); R))/\mathrm{Tor}) \xrightarrow{\cong} \mathrm{End}_\mathcal{C}(L(\bar{H}_*(\Sigma(-); R)))$$

given by $\lambda_3(\alpha) = \sigma_* \circ \alpha \circ \sigma_*^{-1}$. Let

$$f_V\colon L_n(V) \longrightarrow L_m(V)$$

be a natural transformation, where V runs over free R-modules of finite ranks. By Part (1) of Theorem 2.1.3, the composite

$$T_n(V) \xrightarrow{a_1 \otimes \cdots \otimes a_n \mapsto [[a_1, a_2], \ldots, a_n]} L_n(V) \xrightarrow{f_V} L_m(V) \hookrightarrow T_m(V)$$

is trivial if $n \neq m$. Thus

$$\mathrm{Hom}_\mathcal{C}(L_n(\bar{H}_*(\Sigma(-); R)), L_m(\bar{H}_*(\Sigma(-); R))) = 0$$

for $n \neq m$. Moreover there is an isomorphism of rings

$$\mathrm{Hom}_\mathcal{C}(L_n(\bar{H}_*(\Sigma(-); R)), L_n(\bar{H}_*(\Sigma(-); R))) \cong \mathrm{Hom}_{R(S_n)}(\mathrm{Lie}^R(n), \mathrm{Lie}^R(n))$$

by considering X to be n-fold self wedge of S^{2k-1}. It follows that there is an isomorphism of rings

$$\lambda_4 \colon \mathrm{End}_\mathcal{C}(L(\bar{H}_*(\Sigma(-); R))) \xrightarrow{\cong} \prod_{n=1}^{\infty} \mathrm{Hom}_{R(S_n)}(\mathrm{Lie}^R(n), \mathrm{Lie}^R(n)).$$

Let λ be the composite

$$\lambda_4 \circ \lambda_3 \circ \lambda_2 \circ \lambda_1 \colon [\Omega^k\Sigma^2, \Omega^k\Sigma^2] \to \prod_{n=1}^{\infty} \mathrm{Hom}_{R(S_n)}(\mathrm{Lie}^R(n), \mathrm{Lie}^R(n)).$$

The commutativity of Diagram (2.3.1) follows from the facts that the map θ is obtained by restricting to primitives and, for any map $g\colon \Omega Z \to \Omega Z$, there is a commutative diagram

$$\begin{array}{ccc}
\Sigma^k\Omega^{k+1}Z & \xrightarrow{\sigma} & \Omega Z \\
{\scriptstyle \Sigma^k\Omega^k g} \downarrow & & \downarrow {\scriptstyle g} \\
\Sigma^k\Omega^{k+1}Z & \xrightarrow{\sigma} & \Omega Z.
\end{array}$$

2.4. Proof of Theorem 4

This finishes the proof. □

Some terminology is needed before giving the proof. Let M be a manifold and let M_0 be a submanifold of M. Let $F(M, n)$ be the n-th ordered configuration space of M. (See Example 1.2.8 for the definition of $F(M,n)$). Recall that the configuration space $C(M, M_0; X)$ with labels in X is defined to be the quotient space

$$C(M, M_0; X) = \coprod_{n=1}^{\infty} F(M,n) \times_{S_n} X^n / \approx,$$

where $\approx$ is generated by

$$(a_1, ..., a_k; x_1, ..., x_k) \approx (a_1, ..., a_{k-1}; x_1, ..., x_{k-1})$$

if $a_k \in M_0$ or $x_k = *$. The length of a configuration induces a natural filtration of $C(M, M_0; X)$ by the subspaces

$$C_n(M, M_0; X) = \coprod_{k=1}^{n} F(M,k) \times_{S_k} X^k / \approx,$$

$C_0(M, M_0; X) = *$ and $C_1(M, M_0; X) = (M/M_0) \wedge X$ with the k-adic construction of $C(M, M_0; X)$ given by

$$D_n(M, M_0; X) = C_n(M, M_0; X)/C_{n-1}(M, M_0; X) \cong F(M,n)/F(M|M_0, n) \wedge_{S_n} X^{(n)}$$

for each $n \geq 1$ [**6**, pp. 231-239], where $F(M|M_0, n)$ is the subspace of $F(M,n)$ consisting of configurations $(a_1, \ldots, a_n)$ with at least one coordinate in M_0. Recall that if $N \subseteq M$ is a submanifold of codimension 0, and if $N/(N \cap M_0)$ or X is path-connected, then the sequence

(2.4.1) $\quad C(N, N \cup M_0; X) \longrightarrow C(M, M_0; X) \xrightarrow{q} C(M, N \cup M_0; X)$

is a quasifibration, see [**6**, 2.4] or [**26**, Proposition 3.1].

Recall that a map $f: \Sigma Y \to Z$ is called an *evaluation map* if its adjoint map $f': Y \to \Omega Z$ is a (weak) homotopy equivalence. Let $I = [0, 1]$. Consider the commutative diagram

(2.4.2)
$$\begin{array}{ccc}
A(X) & \xrightarrow{\sigma'} & C(M \times I, M_0; X) \\
\downarrow{q'} & \text{pull-back} & \downarrow{q} \\
\bigvee_{i=1}^{2} C(M_i, M_{i,0}; X) & \hookrightarrow & C(M \times I, M_0'; X) \\
\downarrow & & \downarrow{p} \\
* & \longrightarrow & \bigwedge_{i=1}^{2} C(M \times I, M_0^{(i)}; X),
\end{array}$$

where

(1). $M_0 = M \times ([0, 1/5] \cup [4/5, 1])$;
(2). $M_0' = M \times ([0, 1/5] \cup [2/5, 3/5] \cup [4/5, 1])$;

(3). $M_1 = M \times [0, 3/5]$;
(4). $M_{1,0} = M \times ([0, 1/5] \cup [2/5, 3/5])$;
(5). $M_2 = M \times [2/5, 1]$;
(6). $M_{2,0} = M \times ([2/5, 3/5] \cup [4/5, 1])$;
(7). $M_0^{(1)} = M \times ([0, 1/5] \cup [2/5, 1])$;
(8). $M_0^{(2)} = M \times ([0, 3/5] \cup [4/5, 1])$;
(9). The map p is obtained from the canonical quotient map

$$C(M \times I, M_0'; X) \longrightarrow C(M \times I, M_0^{(i)}; X)$$

for $i = 1, 2$.

Following lines in [**6**, 2.4], the map

$$q' : A(X) \to \bigvee_{i=1}^{2} C(M_i, M_{i,0}; X)$$

is a quasifibration. Note that

$$A(X) = C(M \times [0, 3/5], M \times [0, 1/5]; X) \cup C(M \times [2/5, 1], M \times [4/5, 1]; X)$$

is the union of two contractible spaces with the intersection $C(M \times [2/5, 3/5]; X)$. There is a (functorial) filtration-preserving homotopy equivalence

$$\theta : \Sigma C(M \times [2/5, 3/5]; X) \longrightarrow A(X).$$

Observe that the map q is homotopic to the diagonal map of $C(M \times I, M_0; X)$. Thus the pull-back diagram in (2.4.2) is a homotopy pull-back diagram and so

$$\sigma = \sigma' \circ \theta : \Sigma C(M \times [2/5, 3/5]; X) \simeq \Sigma C(M \times I; X) \to C(M \times I, M_0; X) \simeq C(M; \Sigma X)$$

is an evaluation map with the following properties:

1) σ is functorial with respect to X.
2) σ preserves the configuration filtration and so it induces functorial maps

$$\bar{\sigma}_n : \Sigma D_n(M \times I; X) \longrightarrow D_n(M; \Sigma X).$$

3) Let

$$\bar{\psi}_n : D_n(M; \Sigma X) \longrightarrow \bigvee_{i=1}^{n-1} D_i(M; \Sigma X) \wedge D_{n-i}(M; \Sigma X)$$

be the induced map from $p \circ q \simeq \bar{\Delta}_{C(M, \Sigma X)}$ the reduced diagonal. Then the composite

$$\bar{\psi}_n \circ \bar{\sigma}_n : \Sigma D_n(M \times I; X) \longrightarrow \bigvee_{i=1}^{n-1} D_i(M; \Sigma X) \wedge D_{n-i}(M; \Sigma X)$$

is functorially null-homotopic.

In particular, there is an evaluation map

$$\sigma : \Sigma C(\mathbb{R}^2; X) \simeq \Sigma C(I^2; X) \to C(\mathbb{R}; \Sigma X) \simeq J(\Sigma X)$$

with the above three properties.

PROOF OF THEOREM 4. Note that $H_{n-1}(P_n; R) \cong \mathrm{Lie}^R(n)[-1]$ as modules over S_n, see [**7**]. It suffices to show that $\mathfrak{I}_n^R \cong H^{n-1}(B_n; \mathrm{Lie}^R(n)[-1])$. Recall that the unordered configuration space $B(\mathbb{R}^2, n) = F(\mathbb{R}^2, n)/S_n$ is homotopy equivalent to an $(n-1)$-dimensional CW-complex, see for instance [**36, 43**]. The ordered configuration space $F(\mathbb{R}^2, n)$ is S_n-equivariant homotopy equivalent to an $(n-1)$-dimensional CW-complex with free S_n-action. It follows that there exists a chain complex $C = \{C_k\}_{k \geq 0}$ such that

1) Each C_k is a free $\mathbb{Z}(S_n)$-module;
2) $C_k = 0$ for $k \geq n$;
3) There is an S_n-equivariant chain homotopy equivalence

$$\phi \colon C \to S_*(F(\mathbb{R}^2, n)),$$

where $S_*(Y)$ is the singular chain complex of Y.

Given letters $x_1, \ldots, x_n$, let $S^1_{x_j}$ be the circle S^1 labelled by the letter x_j and let $X_n = \bigvee_{j=1}^n S^1_{x_j}$. Let $d_i \colon X_n \to X_{n-1}$ be the map such that $d_i|_{S^1_{x_j}}$ is the inclusion of $S^1_{x_j}$ into X_{n-1} for $j < i$, $d_i|_{S^1_{x_i}} = *$ and $d_i|_{S^1_{x_j}}$ is the inclusion of $S^1_{x_{j-1}}$ into X_{n-1} for $j > i$. Then there is a commutative diagram

$$\begin{array}{ccccc}
\Sigma D_n(\mathbb{R}^2; X_n) & \xrightarrow{\bar{\sigma}_n} & (\Sigma X_n)^{(n)} & \xrightarrow{\bar{\psi}_n} & \bigvee_{j=1}^{n-1} (\Sigma X_n)^{(n)} \\
\downarrow{d_i = D_n(\mathbb{R}^2; d_i)} & & \downarrow{d_i = (\Sigma d_i)^{(n)}} & & \downarrow{d_i = \bigvee_{j=1}^{n-1} (\Sigma d_i)^{(n)}} \\
\Sigma D_n(\mathbb{R}^2; X_{n-1}) & \xrightarrow{\bar{\sigma}_n} & (\Sigma X_{n-1})^{(n)} & \xrightarrow{\bar{\psi}_n} & \bigvee_{j=1}^{n-1} (\Sigma X_{n-1})^{(n)}.
\end{array}$$

for $1 \leq i \leq n$.

Now let $\tilde{\sigma}_n$ be the composite of chain maps

$$\Sigma C \otimes_{\mathbb{Z}(S_n)} (H_1(X_n))^{\otimes n} \longrightarrow S_*(\Sigma D_n(\mathbb{R}^2; X_n)) \xrightarrow{\sigma_{n*}} S_*\left((\Sigma X_n)^{(n)}\right)$$

$$\longrightarrow \bar{S}_*\left((\Sigma X_n)^{(n)}\right) \xrightarrow{\simeq} (H_2(\Sigma X_n))^{\otimes n},$$

where $\bar{S}_*(Y)$ is the reduced singular chain complex of Y. Let

$$\gamma_n^k = \bigoplus_{\sigma \in S_n} H_k(\Sigma^{k-1} S^1_{x_{\sigma(1)}}) \otimes \cdots \otimes H_k(\Sigma^{k-1} S^1_{x_{\sigma(n)}}) \subseteq \left(H_k(\Sigma^{k-1} X_n)\right)^{\otimes n}$$

with S_n-action given by permuting letters $x_1, \ldots, x_n$. By [**39**, Lemma 2.1],

$$\gamma_n^k = \bigcap_{i=1}^n \mathrm{Ker}\left(d_i^{\otimes n} \colon \left(H_k(\Sigma^{k-1} X_n)\right)^{\otimes n} \to \left(H_k(\Sigma^{k-1} X_{n-1})\right)^{\otimes n}\right).$$

Observe that

$$\gamma_n^k \cong \Sigma^{kn} \mathbb{Z}(S_n)[(-1)^k]$$

as modules over the symmetric group S_n.

2.4. PROOF OF THEOREM 4

From the commutative diagram

$$\begin{array}{ccccc}
\Sigma C \otimes_{\mathbb{Z}(S_n)} (H_1(X_n))^{\otimes n} & \xrightarrow{\tilde{\sigma}_n} & (H_2(\Sigma X_n))^{\otimes n} & \xrightarrow{\bar{\psi}_{n*}} & \bigoplus_{i=1}^{n-1}(H_2(\Sigma X_n))^{\otimes n} \\
\downarrow {\scriptstyle \Sigma C \otimes d_i^{\otimes n}} & & \downarrow {\scriptstyle (\Sigma d_i)^{\otimes n}} & & \downarrow {\scriptstyle \oplus (\Sigma d_i)^{\otimes n}} \\
\Sigma C \otimes_{\mathbb{Z}(S_n)} (H_1(X_{n-1}))^{\otimes n} & \xrightarrow{\tilde{\sigma}_n} & (H_2(\Sigma X_{n-1}))^{\otimes n} & \xrightarrow{\bar{\psi}_{n*}} & \bigoplus_{i=1}^{n-1}(H_2(\Sigma X_{n-1}))^{\otimes n},
\end{array}$$

there are induced chain maps

$$\tilde{\sigma}_n \colon \Sigma^{n+1}C[-1] \cong \Sigma C \otimes_{\mathbb{Z}(S_n)} \left(\bigcap_{i=1}^n \mathrm{Ker}(C \otimes d_i^{\otimes n}) \right) \longrightarrow \bigcap_{i=1}^n \mathrm{Ker}((\Sigma d_i)^{\otimes n}) = \gamma_n^2,$$

$$\bar{\psi}_{n*} \colon \gamma_n^2 \longrightarrow \bigoplus_{i=1}^{n-1} \gamma_n^2$$

with the property that $\bar{\psi}_{n*} \circ \tilde{\sigma}_n$ is S_n-equivariant null-homotopic. Since $\bar{\psi}_{n*}$ is induced from the reduced diagonal

$$\bar{\Delta}_* \colon H_*(J(\Sigma X_n)) = T(H_2(\Sigma X_n)) \longrightarrow H_*(J(\Sigma X_n) \wedge J(\Sigma X_n)),$$

we obtain that

$$\bar{\psi}_{n*} = \mathrm{sh} \colon \gamma_n^2 \to \bigoplus_{i=1}^{n-1} \gamma_n^2$$

is the shuffle map. From the fact that $\bar{\psi}_{n*} \circ \tilde{\sigma}_n$ is S_n-equivariant null-homotopic, there is commutative diagram of S_n-modules

(2.4.3)
$$\begin{array}{ccc}
C_{n-1}[-1] & \xrightarrow{\Sigma^{-n-1}\tilde{\sigma}_n} & \Sigma^{-n-1}\gamma_n^2 \\
\downarrow {\scriptstyle \partial_{n-1}^C} & & \downarrow {\scriptstyle \bar{\psi}_{n*} = \mathrm{sh}} \\
C_{n-2}[-1] & \xrightarrow{\phi} & \bigoplus_{i=1}^{n-1} \Sigma^{-n-1}\gamma_n^2.
\end{array}$$

Since $C_k = 0$ for $k \geq n$,

$$\mathrm{Ker}(\partial_{n-1}^C) = H_{n-1}(C)[-1] \cong H_{n-1}(F(\mathbb{R}^2, n), \mathbb{Z}[-1]) = \mathrm{Lie}(n).$$

Moreover the evaluation map $\tilde{\sigma}_n$ induces an isomorphism

$$H_{n-1}(F(\mathbb{R}^2, n); \mathbb{Z}[-1]) \longrightarrow \mathrm{Ker}(\bar{\psi}_{n*} \colon \Sigma^{-n-1}\gamma_n^2 \to \bigoplus_{i=1}^{n-1} \Sigma^{-n-1}\gamma_n^2) = \mathrm{Lie}(n).$$

Thus $\mathrm{Ker}(\partial_{n-1}^C) \cong \mathrm{Ker}(\bar{\psi}_{n*})$ and so $\mathrm{Im}(\partial_{n-1}^C) \simeq \mathrm{Im}(\bar{\psi}_{n*})$ in the stable category of modules over $\mathbb{Z}(S_n)$. From Diagram (2.4.3), it follows that

$$H_{S_n}^{n-1}(C[-1]; M)$$

$$\cong \mathrm{Coker}(\bar{\psi}_{n*}^* \colon \mathrm{Hom}_{S_n}(\bigoplus_{i=1}^{n-1} \Sigma^{-n-1}\gamma_n^2, M) \to \mathrm{Hom}_{S_n}(\Sigma^{-n-1}\gamma_n^2, M))$$

for any $R(S_n)$-module M. In particular,

$$H^{n-1}(B_n; \text{Lie}^R(n)[-1]) \cong H^{n-1}_{S_n}(F(\mathbb{R}^2, n); \text{Lie}^R(n)[-1]) \cong H^{n-1}_{S_n}(C; \text{Lie}^R(n)[-1])$$

$$\cong H^{n-1}_{S_n}(C[-1]; \text{Lie}^R(n))$$

$$\cong \text{Coker}(\text{sh}^* \colon \text{Hom}_{S_n}(\bigoplus_{i=1}^{n-1} \bar{\gamma}_n, \text{Lie}^R(n)) \to \text{Hom}_{S_n}(\bar{\gamma}_n, \text{Lie}^R(n))) \cong \mathfrak{I}_n^R,$$

where $\bar{\gamma}_n = \Sigma^{-2n}\gamma_n^2$ and the last isomorphism is from Proposition 2.1.2 and Diagram (2.1.1). This finishes the proof. □

REMARK 2.4.1. Below are some remarks:
(1). Diagram (2.4.3) is essentially obtained from the cellular filtration of the space $F(\mathbb{R}^2, n)$ by taking cells in the top-two dimensions. So, informally speaking, the shuffle relations are the first type of relations by considering the cellular filtration.
(2). The evaluation maps $\bar{\sigma}_n \colon \Sigma D_n(M \times I; X) \to D_n(M; \Sigma X)$ desuspend. The author observed that a desuspension of $\bar{\sigma}_n$ can be obtained from the pinch map

$$F(M \times I, n) \to F(M \times I, n)/(F((M \times I)|(M \times [0, \epsilon]), n) \cap F((M \times I)|(M \times [1-\epsilon, \epsilon]), n))$$

smashing with $X^{(n)}$ over S_n for small $\epsilon > 0$. It seems that, by considering intersections of the subspaces $F((M \times I)|(M \times [t - \epsilon, t + \epsilon]), n)$, one may get a filtration on $F(M \times I, n)$ and $D_n(M \times I; X)$ built up by wedges of suspensions of $F(M, n)$ and $D_n(M; X)$, respectively. Further study on the homotopy theory of $D_n(M \times I; X \wedge P^m(p^r))$ may be helpful for attacking the Barratt conjecture, where $P^m(p^r)$ is the m-dimensional Moore space with coefficients in $\mathbb{Z}/p^r$.

Bibliography

[1] M. Arkowitz, *Commutators and cup products*, Illinois J. Math. **8** (1964), 571-581.
[2] M. G. Barratt, *On spaces of finite characteristic*, Quart. J. Math. Oxford **11** (1960), 124-136.
[3] J. A. Berrick, F. R. Cohen, Y. L. Wong and J. Wu, *Configurations, Braids and Homotopy Groups*, J. Amer. Math. Soc., to appear.
[4] J. Birman, *Braids, Links and Mapping Class Groups*, Ann. of Math. Studies **82** (1975), Princeton Univ. Press, Princeton, NJ.
[5] H. J. Baues, *Commutator calculus and groups of homotopy classes*, London Mathematical Society Lecture Note Series **50** (1981), Cambridge University Press, Cambridge-New York, ii+160 pp.
[6] C.-F.Bödigheimer, F.R.Cohen and L.Taylor, *On the homology of configuration spaces*, Topology **28** (1989), 111-123.
[7] F.R.Cohen, *Homology of $\Omega^{n+1}S^{n+1}X$ and $C_{n+1}X$ $n > 0$*, Bull.Amer.Math.Soc. **79** (1973), 1236-1241.
[8] F. R. Cohen, *A course in some aspects of classical homotopy theory*, SLNM **1286** (1986), Springer, Berlin, 1-92.
[9] F. Cohen, *On combinatorial group theory in homotopy*, Contemp. Math., **188** (1995), 57-63.
[10] F. R. Cohen, *Fibrations and product decompositions*, Handbook of Algebraic Topology Edited by I. James, Elsevier Science B.V., (1995), 1175-1208.
[11] F. Cohen, *On combinatorial group theory in homotopy, I*, Preprint.
[12] F. Cohen, J. Moore and J. Neisendorfer, *Torsion in homotopy theory*, Ann. Math. **109** (1979), 121-168.
[13] F. R. Cohen, and J. Wu, *On braid groups, free groups, and the loop space of the 2-sphere*, to appear.
[14] E. Curtis, *Simplicial Homotopy Theory*, Advances in Math. **6** (1971), 107-209.
[15] C. De Concini and M. Salvetti, *Cohomology of Artin groups*, Math. Res. Lett. **3** (1996), 293-297.
[16] C. De Concini and M. Salvetti, *Cohomology of Artin groups and Coxeter groups*, Math. Res. Lett. **7** (2000), 213-232.
[17] C. De Concini, C. Procesi and M. Salvetti, *On the equation of degree 6*, preprint.
[18] B. Gray, *On the sphere of origin of infinite families in the homotopy groups of spheres*, Topology **8** (1969), 219-232.
[19] N. Habegger, X.-S. Lin, *Classification of links up to link homotopy*, J. Amer. Math. Soc. **3** (1990), 389-419.
[20] P. J. Hilton, *On the homotopy groups of a union of spheres*, J. London Math. Soc. **30** (1955), 154-172.
[21] I. M. James, *Reduced product spaces*, Ann. Math. **62** (1953), 170-197.
[22] I. M. James, *On the suspension sequence*, Ann. Math. **65** (1957), 74-107.
[23] X.-S. Lin, *Power series expansions and invariants of links,*, 1993 Georgia Topology Conference.
[24] J. P. May, *Simplicial objects in Algebraic Topology*, Math. Studies **11**, van Nostrand (1967).
[25] H. R. Margolis, *Spectra and the Steenrod algebra*, North-Holland Mathematical Library **29**, North-Holland (1983).
[26] D. McDuff, *Configuration spaces of positive and negative particles*, Topology **14** (1975), 91-107.

[27] J. Milnor, *Link groups*, Ann. Math. **78** (1954), 177–195.
[28] J. Milnor, *Isotopy of links*, in *Algebraic Geometry and Topology: A Symposium in Honor of S. Lefschetz*, ed. R. H. Fox, D. C. Spencer, and A. W. Tucker, Princeton Unviersity Press, 1957,pp. 280–306.
[29] J. Milnor, *On the construction FK*, Algebraic Topology-A Student's Guide by J. F. Adams, Cambridge University Press, Cambridge (1972).
[30] J. Milnor and J. Moore, *On the structure of Hopf Algebras*, Ann. Math. **81** (1965), 211-264.
[31] J. A. Neisendorfer, *The exponent of a Moore space*, Algebraic Topology and Algebraic K-theory, Ann. of Math. Studies **113** (1987), 35-71.
[32] G. J. Porter, *Higher order Whitehead products*, Topology **3**(1965), 123-135.
[33] G. J. Porter, *Higher order Whitehead products and Postnikov systems*, Ill. J. Math. **11**(1967), 414-416.
[34] G. J. Porter, *Higher products*, Trans. Amer. Math. Soc. **148**, (1970) 315-345.
[35] C. Reutenauer, *Free Lie Algebras*, Clarendon Press. Oxford, (1993)
[36] M. Salvetti, *The homotopy type of Artin groups*, Math. Res. Lett. **1** (1994), 567-577.
[37] P. S. Selick, *A decomposition of $\pi_*(S^{2p+1};p)$*, Topology **20** (1981), 175-177.
[38] P. S. Selick, *2-primary exponents for the homotopy groups of spheres*, Topology **23** (1984), 97-99.
[39] Paul Selick and Jie Wu, *On natural decompositions of loop suspensions and natural coalgebra decompositions of tensor algebras*, Memoirs AMS **148** (2000), No. 701.
[40] Paul Selick and Jie Wu, *The functor $A^{\min}$ on p-local spaces*, to appear.
[41] Paul Selick and Jie Wu, *Some calculations of $\mathrm{Lie}^{\max}(n)$ for low n*, J. Pure Appl. Algebra, to appear.
[42] S. Smale, *On the topology of algorithms I*, Journal of Complexity **3** (1987), 81-89
[43] J. H. Smith, *Simplicial group models for $\Omega^n \Sigma^n X$*, Israel J. Math. **66** (1989), 330-350.
[44] H. Toda, *On the double suspension E^2*, J. Inst. Polytech. Osaka City Univ., Ser. A **7** (1956), 103-145.
[45] J. Wu, *On combinatorial calculations for the James-Hopf maps*, Topology **37** (1998), 1011-1023.
[46] J. Wu, *On combinatorial descriptions of the homotopy groups of certain spaces*, Math. Proc. Camb. Phil. Soc. **130** (2001), 489-513.
[47] J. Wu, *A braided simplicial group*, Proc. London Math. Soc. **84** (2002), 645-662.
[48] J. Wu, *Homotopy theory of the suspensions of the projective plane*, Memoirs AMS **162** (2003), No. 769.

Editorial Information

To be published in the *Memoirs*, a paper must be correct, new, nontrivial, and significant. Further, it must be well written and of interest to a substantial number of mathematicians. Piecemeal results, such as an inconclusive step toward an unproved major theorem or a minor variation on a known result, are in general not acceptable for publication. Papers appearing in *Memoirs* are generally at least 80 and not more than 200 published pages in length. Papers less than 80 or more than 200 published pages require the approval of the Managing Editor of the Transactions/Memoirs Editorial Board.

As of November 30, 2005, the backlog for this journal was approximately 15 volumes. This estimate is the result of dividing the number of manuscripts for this journal in the Providence office that have not yet gone to the printer on the above date by the average number of monographs per volume over the previous twelve months, reduced by the number of volumes published in four months (the time necessary for preparing a volume for the printer). (There are 6 volumes per year, each containing at least 4 numbers.)

A Consent to Publish and Copyright Agreement is required before a paper will be published in the *Memoirs*. After a paper is accepted for publication, the Providence office will send a Consent to Publish and Copyright Agreement to all authors of the paper. By submitting a paper to the *Memoirs*, authors certify that the results have not been submitted to nor are they under consideration for publication by another journal, conference proceedings, or similar publication.

Information for Authors

Memoirs are printed from camera copy fully prepared by the author. This means that the finished book will look exactly like the copy submitted.

The paper must contain a *descriptive title* and an *abstract* that summarizes the article in language suitable for workers in the general field (algebra, analysis, etc.). The *descriptive title* should be short, but informative; useless or vague phrases such as "some remarks about" or "concerning" should be avoided. The *abstract* should be at least one complete sentence, and at most 300 words. Included with the footnotes to the paper should be the 2000 *Mathematics Subject Classification* representing the primary and secondary subjects of the article. The classifications are accessible from www.ams.org/msc/. The list of classifications is also available in print starting with the 1999 annual index of *Mathematical Reviews*. The Mathematics Subject Classification footnote may be followed by a list of *key words and phrases* describing the subject matter of the article and taken from it. Journal abbreviations used in bibliographies are listed in the latest *Mathematical Reviews* annual index. The series abbreviations are also accessible from www.ams.org/publications/. To help in preparing and verifying references, the AMS offers MR Lookup, a Reference Tool for Linking, at www.ams.org/mrlookup/. When the manuscript is submitted, authors should supply the editor with electronic addresses if available. These will be printed after the postal address at the end of the article.

Electronically prepared manuscripts. The AMS encourages electronically prepared manuscripts, with a strong preference for $\mathcal{AMS}$-LATEX. To this end, the Society has prepared $\mathcal{AMS}$-LATEX author packages for each AMS publication. Author packages include instructions for preparing electronic manuscripts, the *AMS Author Handbook*, samples, and a style file that generates the particular design specifications of that publication series. Though $\mathcal{AMS}$-LATEX is the highly preferred format of TEX, author packages are also available in $\mathcal{AMS}$-TEX.

Authors may retrieve an author package from e-MATH starting from www.ams.org/tex/ or via FTP to ftp.ams.org (login as anonymous, enter username as password, and type cd pub/author-info). The *AMS Author Handbook* and the *Instruction Manual* are available in PDF format following the author packages link from www.ams.org/tex/. The author package can be obtained free of charge by sending email to pub@ams.org (Internet) or from the Publication Division, American Mathematical Society, 201 Charles St., Providence, RI 02904, USA. When requesting an author package, please specify $\mathcal{AMS}$-LaTeX or $\mathcal{AMS}$-TeX, Macintosh or IBM (3.5) format, and the publication in which your paper will appear. Please be sure to include your complete mailing address.

Sending electronic files. After acceptance, the source file(s) should be sent to the Providence office (this includes any TeX source file, any graphics files, and the DVI or PostScript file).

Before sending the source file, be sure you have proofread your paper carefully. The files you send must be the EXACT files used to generate the proof copy that was accepted for publication. For all publications, authors are required to send a printed copy of their paper, which exactly matches the copy approved for publication, along with any graphics that will appear in the paper.

TeX files may be submitted by email, FTP, or on diskette. The DVI file(s) and PostScript files should be submitted only by FTP or on diskette unless they are encoded properly to submit through email. (DVI files are binary and PostScript files tend to be very large.)

Electronically prepared manuscripts can be sent via email to pub-submit@ams.org (Internet). The subject line of the message should include the publication code to identify it as a Memoir. TeX source files, DVI files, and PostScript files can be transferred over the Internet by FTP to the Internet node e-math.ams.org (130.44.1.100).

Electronic graphics. Comprehensive instructions on preparing graphics are available at www.ams.org/jourhtml/graphics.html. A few of the major requirements are given here.

Submit files for graphics as EPS (Encapsulated PostScript) files. This includes graphics originated via a graphics application as well as scanned photographs or other computer-generated images. If this is not possible, TIFF files are acceptable as long as they can be opened in Adobe Photoshop or Illustrator. No matter what method was used to produce the graphic, it is necessary to provide a paper copy to the AMS.

Authors using graphics packages for the creation of electronic art should also avoid the use of any lines thinner than 0.5 points in width. Many graphics packages allow the user to specify a "hairline" for a very thin line. Hairlines often look acceptable when proofed on a typical laser printer. However, when produced on a high-resolution laser imagesetter, hairlines become nearly invisible and will be lost entirely in the final printing process.

Screens should be set to values between 15% and 85%. Screens which fall outside of this range are too light or too dark to print correctly. Variations of screens within a graphic should be no less than 10%.

Inquiries. Any inquiries concerning a paper that has been accepted for publication should be sent directly to the Electronic Prepress Department, American Mathematical Society, 201 Charles St., Providence, RI 02904, USA.

Editors

This journal is designed particularly for long research papers, normally at least 80 pages in length, and groups of cognate papers in pure and applied mathematics. Papers intended for publication in the *Memoirs* should be addressed to one of the following editors. In principle the Memoirs welcomes electronic submissions, and some of the editors, those whose names appear below with an asterisk (*), have indicated that they prefer them. However, editors reserve the right to request hard copies after papers have been submitted electronically. Authors are advised to make preliminary email inquiries to editors about whether they are likely to be able to handle submissions in a particular electronic form.

*Algebra to ALEXANDER KLESHCHEV, Department of Mathematics, University of Oregon, Eugene, OR 97403-1222; email: ams@noether.uoregon.edu

Algebra and its application to MINA TEICHER, Emmy Noether Research Institute for Mathematics, Bar-Ilan University, Ramat-Gan 52900, Israel; email: teicher@macs.biu.ac.il

Algebraic geometry to DAN ABRAMOVICH, Department of Mathematics, Brown University, Box 1917, Providence, RI 02912; email: amsedit@math.brown.edu

*Algebraic number theory to V. KUMAR MURTY, Department of Mathematics, University of Toronto, 100 St. George Street, Toronto, ON M5S 1A1, Canada; email: murty@math.toronto.edu

*Algebraic topology to ALEJANDRO ADEM, Department of Mathematics, University of British Columbia, Room 121, 1984 Mathematics Road, Vancouver, British Columbia, Canada V6T 1Z2; email: adem@math.ubc.ca

Combinatorics to JOHN R. STEMBRIDGE, Department of Mathematics, University of Michigan, Ann Arbor, Michigan 48109-1109; email: jrs@umich.edu

Complex analysis and harmonic analysis to ALEXANDER NAGEL, Department of Mathematics, University of Wisconsin, 480 Lincoln Drive, Madison, WI 53706-1313; email: nagel@math.wisc.edu

*Differential geometry and global analysis to LISA C. JEFFREY, Department of Mathematics, University of Toronto, 100 St. George St., Toronto, ON Canada M5S 3G3; email: jeffrey@math.toronto.edu

Dynamical systems and ergodic theory to AMIE WILKINSON, Department of Mathematics, Northwestern University, 2033 Sheridan Road, Evanston, IL 60208-2730; email: wilkinso@math.northwestern.edu

*Functional analysis and operator algebras to MARIUS DADARLAT, Department of Mathematics, Purdue University, 150 N. University St., West Lafayette, IN 47907-2067; email: mdd@math.purdue.edu

*Geometric analysis to TOBIAS COLDING, Courant Institute, New York University, 251 Mercer St., New York, NY 10012; email: traneditor@cims.nyu.edu

*Geometric analysis to MLADEN BESTVINA, Department of Mathematics, University of Utah, 155 South 1400 East, JWB 233, Salt Lake City, Utah 84112-0090; email: bestvina@math.utah.edu

Harmonic analysis, representation theory, and Lie theory to ROBERT J. STANTON, Department of Mathematics, The Ohio State University, 231 West 18th Avenue, Columbus, OH 43210-1174; email: stanton@math.ohio-state.edu

*Logic to STEFFEN LEMPP, Department of Mathematics, University of Wisconsin, 480 Lincoln Drive, Madison, Wisconsin 53706-1388; email: lempp@math.wisc.edu

*Ordinary differential equations, and applied mathematics to PETER W. BATES, Department of Mathematics, Michigan State University, East Lansing, MI 48824-1027; email: bates@math.msu.edu

*Partial differential equations to GUSTAVO PONCE, Department of Mathematics, South Hall, Room 6607, University of California, Santa Barbara, CA 93106; email: ponce@math.ucsb.edu

*Probability and statistics to KRZYSZTOF BURDZY, Department of Mathematics, University of Washington, Box 354350, Seattle, Washington 98195-4350; email: burdzy@math.washington.edu

*Real analysis and partial differential equations to DANIEL TATARU, Department of Mathematics, University of California, Berkeley, Berkeley, CA 94720; email: tataru@math.berkeley.edu

All other communications to the editors should be addressed to the Managing Editor, ROBERT GURALNICK, Department of Mathematics, University of Southern California, Los Angeles, CA 90089-1113; email: guralnic@math.usc.edu.

Titles in This Series

851 **Jie Wu,** On maps from loop suspensions to loop spaces and the shuffle relations on the Cohen groups, 2006

850 **Siegfried Echterhoff, S. Kaliszewski, John Quigg, and Iain Raeburn,** A categorical approach to imprimitivity theorems for C^*-dynamical systems, 2006

849 **Katsuhiko Kuribayashi, Mamoru Mimura, and Tetsu Nishimoto,** Twisted tensor products related to the cohomology of the classifying spaces of loop groups, 2006

848 **Bob Oliver,** Equivalences of classifying spaces completed at the prime two, 2006

847 **Eric T. Sawyer and Richard L. Wheeden,** Hölder continuity of weak solutions to subelliptic equations with rough coefficients, 2006

846 **Victor Beresnevich, Detta Dickinson, and Sanju Velani,** Measure theoretic laws for lim-sup sets, 2006

845 **Ehud Friedgut, Vojtech Rödl, Andrzej Ruciński, and Prasad V. Tetali,** A Sharp threshold for random graphs with a monochromatic triangle in every edge coloring, 2006

844 **Amadeu Delshams, Rafael de la Llave, and Tere M. Seara,** A geometric mechanism for diffusion in Hamiltonian systems overcoming the large gap problem: Heuristics and rigorous verification on a model, 2006

843 **Denis V. Osin,** Relatively hyperbolic groups: Intrinsic geometry, algebraic properties, and algorithmic problems, 2006

842 **David P. Blecher and Vrej Zarikian,** The calculus of one-sided M-ideals and multipliers in operator spaces, 2006

841 **Enrique Artal Bartolo, Pierrette Cassou-Noguès, Ignacio Luengo, and Alejandro Melle Hernández,** Quasi-ordinary power series and their zeta functions, 2005

840 **Sławomir Kołodziej,** The complex Monge-Ampère equation and pluripotential theory, 2005

839 **Mihai Ciucu,** A random tiling model for two dimensional electrostatics, 2005

838 **V. Jurdjevic,** Integrable Hamiltonian systems on complex Lie groups, 2005

837 **Joseph A. Ball and Victor Vinnikov,** Lax-Phillips scattering and conservative linear systems: A Cuntz-algebra multidimensional setting, 2005

836 **H. G. Dales and A. T.-M. Lau,** The second duals of Beurling algbras, 2005

835 **Kiyoshi Igusa,** Higher complex torsion and the framing principle, 2005

834 **Kenichi Ohshika,** Kleinian groups which are limits of geometrically finite groups, 2005

833 **Greg Hjorth and Alexander S. Kechris,** Rigidity theorems for actions of product groups and countable Borel equivalence relations, 2005

832 **Lee Klingler and Lawrence S. Levy,** Representation type of commutative Noetherian rings III: Global wildness and tameness, 2005

831 **K. R. Goodearl and F. Wehrung,** The complete dimension theory of partially ordered systems with equivalence and orthogonality, 2005

830 **Jason Fulman, Peter M. Neumann, and Cheryl E. Praeger,** A generating function approach to the enumeration of matrices in classical groups over finite fields, 2005

829 **S. G. Bobkov and B. Zegarlinski,** Entropy bounds and isoperimetry, 2005

828 **Joel Berman and Paweł M. Idziak,** Generative complexity in algebra, 2005

827 **Trevor A. Welsh,** Fermionic expressions for minimal model Virasoro characters, 2005

826 **Guy Métivier and Kevin Zumbrun,** Large viscous boundary layers for noncharacteristic nonlinear hyperbolic problems, 2005

825 **Yaozhong Hu,** Integral transformations and anticipative calculus for fractional Brownian motions, 2005

824 **Luen-Chau Li and Serge Parmentier,** On dynamical Poisson groupoids I, 2005

TITLES IN THIS SERIES

823 **Claus Mokler,** An analogue of a reductive algebraic monoid whose unit group is a Kac-Moody group, 2005

822 **Stefano Pigola, Marco Rigoli, and Alberto G. Setti,** Maximum principles on Riemannian manifolds and applications, 2005

821 **Nicole Bopp and Hubert Rubenthaler,** Local zeta functions attached to the minimal spherical series for a class of symmetric spaces, 2005

820 **Vadim A. Kaimanovich and Mikhail Lyubich,** Conformal and harmonic measures on laminations associated with rational maps, 2005

819 **F. Andreatta and E. Z. Goren,** Hilbert modular forms: Mod p and p-adic aspects, 2005

818 **Tom De Medts,** An algebraic structure for Moufang quadrangles, 2005

817 **Javier Fernández de Bobadilla,** Moduli spaces of polynomials in two variables, 2005

816 **Francis Clarke,** Necessary conditions in dynamic optimization, 2005

815 **Martin Bendersky and Donald M. Davis,** V_1-periodic homotopy groups of $SO(n)$, 2004

814 **Johannes Huebschmann,** Kähler spaces, nilpotent orbits, and singular reduction, 2004

813 **Jeff Groah and Blake Temple,** Shock-wave solutions of the Einstein equations with perfect fluid sources: Existence and consistency by a locally inertial Glimm scheme, 2004

812 **Richard D. Canary and Darryl McCullough,** Homotopy equivalences of 3-manifolds and deformation theory of Kleinian groups, 2004

811 **Ottmar Loos and Erhard Neher,** Locally finite root systems, 2004

810 **W. N. Everitt and L. Markus,** Infinite dimensional complex symplectic spaces, 2004

809 **J. T. Cox, D. A. Dawson, and A. Greven,** Mutually catalytic super branching random walks: Large finite systems and renormalization analysis, 2004

808 **Hagen Meltzer,** Exceptional vector bundles, tilting sheaves and tilting complexes for weighted projective lines, 2004

807 **Carlos A. Cabrelli, Christopher Heil, and Ursula M. Molter,** Self-similarity and multiwavelets in higher dimensions, 2004

806 **Spiros A. Argyros and Andreas Tolias,** Methods in the theory of hereditarily indecomposable Banach spaces, 2004

805 **Philip L. Bowers and Kenneth Stephenson,** Uniformizing dessins and Belyĭ maps via circle packing, 2004

804 **A. Yu Ol'shanskii and M. V. Sapir,** The conjugacy problem and Higman embeddings, 2004

803 **Michael Field and Matthew Nicol,** Ergodic theory of equivariant diffeomorphisms: Markov partitions and stable ergodicity, 2004

802 **Martin W. Liebeck and Gary M. Seitz,** The maximal subgroups of positive dimension in exceptional algebraic groups, 2004

801 **Fabio Ancona and Andrea Marson,** Well-posedness for general 2×2 systems of conservation law, 2004

800 **V. Poénaru and C. Tanas,** Equivariant, almost-arborescent representation of open simply-connected 3-manifolds; A finiteness result, 2004

799 **Barry Mazur and Karl Rubin,** Kolyvagin systems, 2004

798 **Benoît Mselati,** Classification and probabilistic representation of the positive solutions of a semilinear elliptic equation, 2004

797 **Ola Bratteli, Palle E. T. Jorgensen, and Vasyl' Ostrovs'kyĭ,** Representation theory and numerical AF-invariants, 2004

For a complete list of titles in this series, visit the
AMS Bookstore at **www.ams.org/bookstore/**.